brilliant!

brilliant!

shuji nakamura
and the Revolution in
Lighting Technology

Bob Johnstone

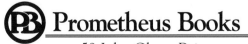 **Prometheus Books**

59 John Glenn Drive
Amherst, New York 14228-2197

Published 2007 by Prometheus Books

Inquiries should be addressed to
Prometheus Books
59 John Glenn Drive
Amherst, New York 14228–2197
VOICE: 716–691–0133, ext. 207
FAX: 716–564–2711
WWW.PROMETHEUSBOOKS.COM

11 10 09 08 07   5 4 3 2 1

Library of Congress Cataloging-in-Publication Data

Johnstone, Bob.
    Brilliant! : Shuji Nakamura and the revolution in lighting technology / by Bob Johnstone.
        p. cm.
    Includes bibliographical references and index.
    ISBN 978–1–59102–462–0
    1. Nakamura, Shuji, 1954– 2. Light emitting diodes. 3. Blue light. 4. Inventors—Japan—Biography. 5. Electric engineers—Japan—Biography. 6. Lighting—History. I. Title. II. Title: Shuji Nakamura and the revolution in lighting technology.

TK7871.89.L53J65 2007
621.32092—dc22
[B]   2007003579

Printed in the United States of America on acid-free paper

To Victor McElheny

CONTENTS

8   CONTENTS

INTRODUCTION

The eminent historian of technology Tom Hughes employs a military metaphor to describe problems that remain to be solved as technological systems expand. He calls them "reverse salients." Innovation advances across a broad front, leaving reverse salients sticking out behind. These pesky pockets of resistance often defy efforts to mop them up. They bring progress elsewhere grinding to a halt. For obvious reasons, reverse salients attract ambitious inventors. A classic case of a reverse salient is the bright blue light emitting diode. LEDs—tiny specks of semiconductor material that shine when hooked up to a voltage—were invented in 1962. Thirty years later, researchers had made considerable progress in refining the technology. From its humble origins as a dull red on/off indicator, the LED had evolved to become, among other things, the bright red brake lights on cars. Along the way, other colors had also emerged: amber, yellow, and yellow-green. But the light emitting diode was still a long way from fulfilling the destiny that the device's pioneers had predicted for it. That is, to replace every other form of lighting, including Edison's lightbulb and the fluorescent tube, in our homes, offices, and everywhere else besides.

Such a large-scale replacement would be a very big deal, for all sorts of reasons. First and foremost, in terms of reduced energy consumption. Symbol of innovation though it may be, the incandescent lightbulb is in fact a throwback to the nineteenth century. Edison's best-known invention is a gross object made from gas, glass, and brass. Suspended within the glass is a flimsy, very Victorian-looking metal contraption called a filament. Lightbulbs guzzle energy, wasting 95 percent of their output in the form of heat. And, as we all know from bitter experience, lightbulbs are irritatingly prone—plink!—to burn out after only a few months. One in three bulbs needs replacing every year.

Light emitting diodes, by contrast, are solid-state devices made up of layers of semiconductor material just a few atoms thick. They are arguably the world's first ubiquitous nanotechnology. LEDs are cool, both literally—touch one and see—and metaphorically, too, in terms of what you can do with them. They consume 80 percent less energy than incandescents. They are mechanically robust, almost unbreakable. They last for up to a hundred thousand hours, over a decade, effectively forever (because you don't leave lights turned on all the time). And like old soldiers, LEDs do not die; they gracefully fade away.

Since lighting accounts for around a quarter of electricity usage, replacing conventional lights with LEDs would dramatically cut our energy consumption. In the United States, by far the world's largest user of electricity, energy consumption would decline by almost 30 percent. By switching to solid-state lighting, consumers might expect to save $125 billion over the next twenty years. That does not include many more billions of dollars in construction costs that could be avoided as a result of not having to build new power plants. (In the United States alone, 135 new coal-fired plants are currently on the drawing board.) Fewer fossil-fuel electric plants would mean a reduction in carbon emissions of hundreds of millions of tons. In addition to this, LEDs are intrinsically environment friendly. They contain no toxic substances like mercury, which is used in fluorescent tubes, and thus do not threaten landfills, groundwater, and ultimately our health.

Light emitting diodes would generate economic growth, giving birth to a whole new industry. The products of this industry would do far more than just replace existing lights. As Herbert Kroemer—the Nobel laureate from whose fundamental insights modern LEDs derive their efficiency in converting electricity into light—instructs, a breakthrough technology creates its own applications. Witness the transistor, the most significant application of which was not a replacement for tubes in radios and TVs but the sine qua non of the personal computer, an application that no one imagined initially. LEDs, like transistors, are semiconductors. Their performance keeps going up even as their price keeps coming down. Haitz's law, the LED equivalent of Moore's law, states that each decade since the first LED appeared, device performance—in this case, the amount of light output—has increased by twenty times, while the device price has fallen by ten times.

Solid-state lighting couples naturally with digital devices like microprocessors to spawn a host of innovative applications. For example, you might be able to adjust the color of the lighting in your living room to suit the occasion, the time of day, or your mood. (Northern Lights? Sunset in Maui? Not a problem.) This ability to do new and previously impossible things may trigger the mass adoption of novel household lighting systems. The monster market will not be fancy tricks, however, but plain old illumination. That is, replacing incandescents, halogens, and ultimately fluorescents as the lighting technology of choice in every single socket in the world. For many people, the "killer app" of solid-state lighting may simply be tens of dollars off the monthly electricity bill.

For LEDs to fulfill their destiny, however, for this unprecedented paradigm shift to occur, one breakthrough remained to be made. In order to produce useful white light, you need—or so people thought in the early nineties—all three primary colors: not just bright red, but also bright blue and bright green. Over three decades, elite scientists at some of the world's top universities and corporate research centers strove to solve this problem. Despite all efforts, no one was able to devise a solution that would rectify the reverse salient, thus enabling technological progress toward solid-state lighting to resume. You could do blue LEDs, it was true, but they were very far from being bright, and seemed destined to remain so.

Then, on November 29, 1993, came an announcement that astonished the world (or at least that part of the world that knew enough to be astonished). A chink in Nature's armor had been found! A bright blue light emitting diode had been invented! The days of lone inventors like Edison were supposedly long gone. (It took three men to create the transistor.) Now there was proof to the contrary. The finder of the chink in Nature's armor was an individual inventor named Shuji Nakamura. His little lamp was about a hundred times brighter than previous blue LEDs. It was, quite literally, brilliant. The reverse salient of solid-state lighting was history.

Initial reaction to news of the discovery of the bright blue LED was one of disbelief. Who was this unknown inventor, this presumptuous upstart without so much as a PhD to his name, this latter-day alchemist who claimed to have turned base metal into light emitting gold? Shuji Nakamura, it turned out, was a humble thirty-six-year-old Japanese electrical engineer. He had been working by himself, in almost total isolation—personal, professional, and geographical. The company that employed him was called Nichia Chemical Industries. It was a small, utterly obscure outfit based on the island of Shikoku, way down in the boondocks of rural Japan. In American terms, it was as if the announcement had come from hillbilly Appalachia.

Since then, having been held back for so long, LED technology has exploded into all sorts of applications. In fewer than fifteen years, Nichia's tiny lights—along with similar devices made by rival firms—have become ubiquitous. Today, you carry them in your pocket, lighting the keypad of your cell phone, backlighting its display, and providing the flash for its camera. They are on your desktop, in your computer, and in its screen, mouse, and flash drive. They are in the jewelry store, illuminating the display cabinets. They are on the roadside, in the form of pointillist traffic signals and vast, full-color electronic billboards. Tomorrow, Nakamura-style high-brightness LEDs will be found in such applications as headlights in cars, shelf lights in supermarkets, reading lights (and every other kind of light) in airplanes, and, most important, in our homes.

The efficiency with which LEDs convert electricity into light, already high, continues to increase. Soon it will reach that of fluorescent tubes. The main problem now is cost, and cost has never stood in the way of semiconductors. As they come pounding down the experience curve, semiconductors are like a steamroller, flattening all obstacles in their path. The solid-state lighting revolution is happening now, faster than anyone imagined. Lighting industry stalwarts who used to say, "It shows promise—but not in my lifetime," or, "Let's wait until it reaches the crossover point," are now frantically scrambling to climb aboard the LED bus. Light emitting diodes are already the lighting technology of choice in some high-end California houses. By 2010, as the performance of the devices continues to improve and their prices to plummet, LEDs will be replacing incandescent lightbulbs in homes across the United States and fluorescent tubes in offices and commercial spaces. By 2020 the tiny lights will likely have superseded all conventional forms of illumination, with the possible exception of searchlights.

✳   ✳   ✳

It behooves scientists to be modest about their discoveries. "If I have seen further," Isaac Newton said, "it is by standing on the shoulders of giants." As Shuji Nakamura himself is the first to admit, his initial bright blue breakthrough, though based on his own innovations and a decade's worth of blood, sweat, and tears, was built on the work of others. But then, from his platform atop others' shoulders, Shuji took off. It was as if he had rockets in his feet like Mighty Atom, his boyhood comic-book superhero.

Over the next six years, from 1993 to 1999, Nakamura notched one key breakthrough after another. Following bright blue LEDs came bright emerald green LEDs (thus completing the trinity of primary colors), blue-violet lasers (thus enabling next-generation, high-definition DVD players), and ever-brighter blue LEDs (thus producing, by adding a yellow phosphor, a simpler, hence cheaper, white light than the combination of red-blue-green). And all of these inventions were undeniable; you could actually *see* them with your own eyes.

At international conference after international conference, Shuji the Showman would blind audiences by shining his latest little lights in their faces. Cheekily, he would use the world's first blue laser as a pointer in his presentations. Such antics infuriated his rivals, because Shuji would always be at least one step ahead of them. Some accused Nakamura of being an egomaniac. In fact, he was just being a great salesman, à la Steve Jobs, advertising his company's products.

Nakamura appeared—to use an entirely appropriate expression—from out of the blue. "Nobody had ever met him, had ever heard of the guy," marveled Gerald Stringfellow, formerly of Hewlett-Packard, now a distinguished professor of materials science and engineering at the University of Utah and coauthor of a book on LEDs. "Then suddenly, he came out of nowhere, and just one thing after another fell in front of him. . . . And I have been absolutely flabbergasted that one person could make such a big contribution, could be such a leader for such a long time."

"Never during the time that I am aware of," said Fernando Ponce, formerly a researcher at Xerox's Palo Alto Research Center and now a professor of physics at Arizona State University, "have we seen somebody maintain such a huge lead, and the rest of the world not being able to compete with him. . . . It's a wonderful demonstration of what a brilliant engineer can do."

Not even a US government-sponsored consortium of US companies—including Xerox and Hewlett-Packard—that was thrown together in a misguided attempt to replicate the supposed Japanese cooperative-style of technology development could catch him.

"Nakamura put together a string of achievements," wrote Glenn Zorpette in *Scientific American*, "that for genius and sheer improbability is *as impressive as any other accomplishment in the history of semiconductor research* [my emphasis]. And it is all documented in a trail of literature that is almost as stunning. Between 1991 and 1999 he authored or co-authored 146 technical papers, six books and 10 book chapters."

Nakamura's breakthroughs laid the foundations for a whole new industry that today employs tens of thousands in Japan, North America, Germany, Taiwan, South Korea, China, and elsewhere. Participants include a few big names, plus a host of brash, entrepreneur-driven start-ups.

Along the way Shuji has garnered most of the honors and awards going, both at home and abroad. In 2002 alone, they included Japan's prestigious Takeda Award and the Franklin Institute's Medal in Engineering. (Named for Ben, the Franklin awards are widely regarded as the American Nobels.) A prediction by Thomson Scientific based on a citation count of his work puts Nakamura on the short list for the Nobel Prize for Physics. The middle-of-the-night phone call from Stockholm may not come this year, or next, but as Shuji's tiny electronic progeny conquer ever-larger markets with an ever-greater effect on our lives and the environment, there can be little doubt that, sooner or later, it will come. To win a Nobel Prize, according to Nobel Museum director Svante Linqvist, you do not have to be a genius: what you need most of all is courage. In order to come up with his breakthrough Shuji had to defy the conventional wisdom of his peers and ignore the explicit instructions of his boss. He does not lack guts.

*   *   *

In 1998, after five years of setting the pace, Shuji announced at a plenary talk he gave in Strasbourg, France, that LED research was over. An exaggeration, as we shall see, but one with a kernel of truth. Most of the big breakthroughs had been made, many of them by Shuji and his group at Nichia. But the Nakamura saga was by no means over. Indeed, its most dramatic episodes were about to begin. In January 2000 came the sensational news that Shuji had done the unthinkable. He had left Nichia, the company where he had worked for twenty years, and his beloved native Shikoku—something he had vowed he would never do—to come to the United States and take up a chair as a professor at the University of California at Santa Barbara. In Japan, still very much the land of lifelong employment at one company, it was a shocking thing to do.

At first, the parting of ways seemed amicable. Shuji had gone as far as he could go with the technology; now he was leaving to pursue a more scientific vocation. In fact, as the truth of the matter emerged, the split was anything but friendly. In December 2000 Nichia sued Nakamura, accusing him of leaking trade secrets to Cree, a rival US firm. Shuji countersued, claiming that his old employer had not rewarded him commensurately for his efforts in laying the foundations for what had by then for Nichia become a business worth well over a billion dollars.

It subsequently emerged that while Nichia's founder had supported Nakamura's bold challenge to develop a bright blue LED, his successor as president had done everything in his power to sideline his company's most valuable human resource. In an act of breathtaking ingratitude, he succeeded in chasing away the goose that had laid the golden eggs.

The court battles dragged on for years. In the interim, via public lectures and op-ed pieces in Japanese newspapers and magazines, Shuji became a vocal critic of the Japanese education, employment, and legal systems. Eventually, the case culminated in victory not only for Shuji, but also for the rights of the creative individual in a corporate context. Never again would a Japanese corporation be able to take for granted the efforts of its most innovative employees. Nakamura emerged from this epic legal struggle as a folk hero among the downtrodden salarymen of his native land.

*   *   *

The semiconductor material enabling the revolution in lighting is a compound called gallium nitride. Some experts claim that gallium nitride is the most important new material since silicon. When Shuji announced his breakthrough back in 1993, all the researchers working on gallium nitride could have fit into a minibus. Twelve years later, a survey listed no fewer than 626 companies, universities, and research institutes worldwide actively involved with gallium nitride, of which 232 were companies. Not all gallium nitride researchers are working on light emission, however. Shuji's breakthrough has also cracked open another potentially huge set of new markets.

Umesh Mishra, Nakamura's colleague at UC Santa Barbara, predicts that gallium nitride will also have a disruptive effect in the area of power-related devices. Such gadgets will improve the energy efficiency of delivering power, just like solid-state lighting improves the energy efficiency of delivering illumination. There will be applications for gallium nitride in broadband wireless networks, mobile computing, and hybrid vehicles—anywhere, in short, where you have to convert electricity from AC to DC, or from one voltage to another. For example, in transformers, the ugly "bricks on the wall" that terminate at the ends of the power cords of so many appliances. Such things are pervasive in our lives, and they can keep getting smaller, lighter, and more efficient. Mishra believes that a huge second wave of gallium nitride applications is on its way.

* * *

Shuji Nakamura's invention of the bright blue light emitting diode and the explosion of entrepreneurial activity that has ensued is more than just a tale of derring-do in the semiconductor industry. The story also contains important lessons for society on how to foster and reward innovation, and hence build new industry and create wealth. Not, of course, that it would be possible to replicate the highly contingent circumstances that led to Nakamura making his breakthrough. Nonetheless, Shuji's success basically boils down to two things. One, a talented, highly motivated researcher is given free rein by, two, a chief executive willing to gamble big on his abilities. Though Nakamura had to overcome more than his fair share of obstacles, he never had to contend with the risk-averse bureaucracy that stifles creativity at so many big corporations.

"What I have managed to achieve," Nakamura has written, "shows that anybody with relatively little experience in a field, with no big money and no collaborations with universities or other companies, can achieve considerable research success alone when he tries a new research area without being obsessed by conventional ideas and wisdom."

Though Nakamura's breakthroughs took place at a small Japanese company, and his research flourishes today at an entrepreneur-friendly American university, neither Japan nor the United States have much reason to feel complacent about these successes.

- Americans should ask themselves why their federal government has done so little to promote the crucial energy-saving technology of LEDs. This is in stark contrast to other nations, especially Asian countries, China in particular, which have identified solid-state lighting as a national priority and are supporting it accordingly. Also why, despite being the source of much of the advanced technology that is driving the solid-state lighting revolution, the United States has been so slow to adopt that technology. Compared to their European counterparts, US lighting companies seem reluctant to embrace LEDs and progress has been, at best, gradual.

In addition, America's much-vaunted venture capitalists do not emerge from this story well. For the most part, angel investors—that is to say, individuals—have driven the creation of the new industry. Venture capitalists turned a blind eye to real value-adding solid-state lighting start-ups, instead flocking like sheep to back the worthless wet dreams of the dot-com bubble.

- Japanese should ask themselves why, despite the facts that the bright blue light emitting diode originated in Japan and that two of the "Big Five"* manufacturers of gallium nitride LEDs are Japanese, essentially none of the huge surge of related entrepreneurial activity has taken place in Japan. Almost all the solid-state lighting start-ups that have popped up over the past few years have been American, Asian, or, interestingly enough, Canadian. Though the regulatory environment in Japan is no longer as hostile to entrepreneurs as it once was, the Japanese still have much to learn about how to foster risk taking and wealth creation.

- For their part, Europeans should stop patting themselves on the back for designing cool new products that incorporate LEDs. Instead, they should ask themselves why, despite the fact that several of the actors in this drama are European by birth, it is possible to write a book such as this essentially without mentioning Europe. Indeed, it is only a slight exaggeration to say that, from a technological and entrepreneurial point of view regarding LEDs, with the honorable exception of Germany's Osram Opto, Europe hardly exists.

✳   ✳   ✳

---

* The Big Five consist of Nichia, Toyoda Gosei, Cree, LumiLEDs, and Osram Opto. In addition, Epistar, the leading Taiwanese LED maker, should probably now be included.

In more than twenty years of writing about high tech, I must have met and interviewed thousands of scientists and engineers. Many of them had remarkable tales to tell. But in my experience, nothing matches the extraordinary story of the brilliant Japanese engineer Shuji Nakamura. Indeed, for sheer unlikeliness, nothing even comes close.

This book recounts Shuji Nakamura's inspirational story, much of it told here for the first time. It is based on face-to-face interviews with Shuji himself, whom I first met in 1994 soon after he had made his presence felt and whom I have since interviewed on many occasions, both at Nichia and at UC Santa Barbara; also, on numerous interviews with his colleagues, contemporaries, peers, rivals, and students; and on his own writings, most notably his Japanese-language autobiography, *Ikari no Bureikusuru* [*Breakthrough with Anger*], which my wife and I translated into English for this project.

In addition to chronicling the life of a brilliant engineer, this book also investigates some of the many and various ways Shuji's brilliant inventions and their derivatives are making their way into our lives, opening up new possibilities, changing forever the way we see our world. Nakamura blew open the floodgates, allowing the creative juices of innovators of all stripes to flow. They include entrepreneurs, philanthropists, designers, architects, and artists who are using solid-state lighting to create new products and eye-catching artwork. For example, navigation lights, emergency lighting, street signs, bus stops, pocket-sized projectors, water purifiers, adhesive curing systems, portable anthrax detectors, reading lights, light sculptures, ice cubes that light up, buildings that iridesce like squid, and affordable off-grid lighting systems for poor villagers in developing countries, to name but a few. Finally, this book looks at the momentous ongoing revolution that is taking place in lighting. Based on interviews with pioneers in the field, it speculates on how LEDs will replace lightbulbs and fluorescent tubes, and when and where the change will take place.

Shuji Nakamura's high-brightness LEDs are a disruptive technology, happening in real time, right in front of our eyes. The story of their development, commercialization, and application is, ultimately, one of human creativity in all its forms and at its most dynamic.

Part One

OUT OF THE BLUE

CHAPTER 1

S huji Nakamura would never have made his brilliant, world-changing, bright blue breakthrough had he not while still a student done something that was, by Japanese standards, absolutely outrageous.

He got married.

And then, to make matters worse, he and his wife, Hiroko, had a baby.

The year was 1979 and, despite the effects of the second oil crisis, the Japanese electronics industry was just entering a boom that would last more than a decade. As a twenty-five-year-old graduate student with a master's degree in electrical engineering from Tokushima University, Shuji naturally expected that one of Japan's monster consumer appliance manufacturers would swallow him in its maw. Joining a big company would guarantee him lifelong employment as a salaryman. But the likes of Sony tend not to recruit married grads. The corporate dorms where such firms house their new hires are not suitable for couples, especially not couples with kids.

His marital handicap notwithstanding, Shuji did try to get a job at a big company. At Matsushita, he had the advantage of a recommendation from his university. But in his entrance exam he made the mistake of discussing the theoretical aspects of his thesis research. This was not a good idea. "We don't need theoreticians," the giant Osaka-based firm's recruiters told him as they rejected his application. (Matsushita was ever thus: when Konosuke Matsushita's scientists presented their company's founder with some promising new technology, the old man would typically ask just one question: "Will it make money?")

Nakamura told his thesis adviser at Tokushima, Osamu Tada, what had happened. His advising professor laughed at him: "Companies are in the business of making things—it's not surprising you failed if you wrote about theory."

At Kyocera, despite the lack of a recommendation, Shuji did better. Acting on his professor's advice he emphasized in his entrance exam possible practical applications of his work. Kyocera is known for recruiting employees from second-division technical schools—of which Tokushima was undoubtedly one—with the belief that, give them a chance and they will work that much harder. Shuji was interviewed by Kazuo Inamori, one of Japan's handful of successful high-tech entrepreneurs. "If you join our company, what would you like to do?" Kyocera's chief executive wanted to know. "Wherever you assign me," Shuji replied enthusiastically, "even if it's sales or accounting, I'm confident I can do well if you give me the opportunity."

This answer was evidently satisfactory, because Nakamura made it through to the next round of interviews. He was summoned to appear at the company's headquarters in Kyoto by eight o'clock the next morning. The journey presented the young man with logistical problems. Shuji was a country boy, unused to making long trips. Prior to his job-hunting sallies to the mainland he had rarely left Shikoku, the island of his birth.

To get to Kyocera he had to take the ferry from his home in Tokushima to Osaka, then get the train to Kyoto. But Nakamura underestimated the time the boat would take, and he was late arriving in Osaka. Fretting that he would not be on time, Shuji took a taxi all the way to the company. It was an expensive ride: every time the meter ticked over, his heart leapt. "I had never had such a hard time," he recalled many years later.

Shuji subsequently went back to Kyocera a third time. There, he was informed that the company had decided to offer him a job. He was also told the reason for the decision: he was a strange person. "Strange in what way?" Shuji wanted to know. "Because you are married," came the reply. The young man was, somewhat perversely, pleased to hear that his prospective employer considered him a bit of an oddball. But he was also relieved. Until then he had been an impecunious student, living with his wife's parents. Now, with a job in the big city, he would be able to earn a salary and support his family.

This was in February. But as April 1, the day on which Nakamura was due to report for work in Kyoto neared, he started having second thoughts. Shuji recalled a high school trip to Tokyo. For a teenage boy from a tiny country town, the metropolis had been an almost terrifying experience. All those crowds, the jam-packed rush-hour commuter trains. Just thinking about it made him feel sick.

Shuji had always been a nature lover. He treasured the ocean and the mountains, with which Shikoku is particularly well endowed. Now, as he contemplated his baby daughter Hitomi sleeping peacefully on the tatami mat in front of him, he determined that he did not want her to grow up in a big city. Had he still been single it might have been different. But not with a family: a big city was no place to raise children. At the same time, Shuji really wanted to work in a proper research laboratory at a major company. Characteristically for a Japanese man, he had not mentioned to his wife, who taught at a kindergarten attached to Tokushima University, that Kyocera had offered him a job. He had intended to tell Hiroko once everything was finalized.

Torn between two options, Shuji went to see Professor Tada and asked him what he should do—stay in Tokushima, or head for the mainland? "If you join a big company like Kyocera," Tada told his student, "you'll be a salaryman all your life, shunted around from one posting to the next. That's tough on a family. Your wife has a proper job, her family is here, so why don't you stay in Tokushima?" Having given this advice, Tada immediately pointed out the obvious drawback of such a decision. In Tokushima, there was nowhere for an electrical engineer to work. If Shuji elected to stay on the island, he might have a happy family life, but he would have to give up any notion of pursuing a career in his chosen field.

Following this conversation, Shuji agonized for two weeks over what to do. The choice was simple: family or career. Still, he could not bring himself to discuss the dilemma with Hiroko. The decision would be his and his alone. In the end, he decided to put his wife and family first—they would stay in Tokushima. He went back to see his professor to explain his decision. "I'm not going to join Kyocera," Nakamura told Tada. "I'm giving up my dream of doing research in a big company. Please help me to find a job—it doesn't matter what it is, I'll take whatever I can get."

Having understood that the young man had indeed made up his mind, Tada considered the problem. Then he replied: "I may be able to help. The president of Nichia Chemical is a good friend of mine. He's from Anan, the same hometown as me. I might be able to introduce him to you. Nichia's the only place I can think of—will that be all right?"

Nakamura had been living in Tokushima for six years, studying for his undergraduate and master's degrees. During that time he had never once heard of a company called Nichia Chemical. Nor was he altogether sure where Anan was. Nonetheless, he had little option but to accept Tada's offer of an introduction to the president of this obscure rural firm. Sensing his unease, Tada reassured him. "Don't worry, although it's a chemical company, they have a good measurements department, and they excel in analysis technology. I hear they're planning to make a move into the electronic device field. They may not be as good as Philips, but they're one of the world's leading manufacturers of phosphors."

Nakamura was not entirely sure that it was proper to compare Europe's top consumer appliance maker with a company whose name he had never heard, and that was based in a town whose whereabouts were unknown to him. He was also not convinced that, simply because the president was a good friend of his professor, Nichia would be prepared to hire him, just like that. After all, as Tada admitted, he had never introduced any of his students to the company before. "You may be number one in engineering here at the university," his professor conceded, "but I'm not sure whether Nichia will take you. We'll just have to go there and see."

So, one fine morning in late March, Nakamura met up with his professor at the university, and in the latter's car the pair headed off for the little town of Anan, some twenty miles to the south. As they drove down the narrow back roads, Nakamura began to feel uneasy. This was the sticks: could there really be a company located in such an out-of-the-way place? About an hour after leaving the university, they finally arrived at the firm's premises, located in the midst of green fields, surrounded by what appeared to be a pine forest. On the gate hung a wooden sign that read "Nichia Chemical Industries Ltd." For the most part the site consisted of single-story buildings roofed with galvanized iron. The place looked a bit like an old army barracks.

Nakamura got out of the car, then gasped as the stench of bad eggs hit him. It was sulfur, such as you might smell at a volcanic hot spring. Nichia used sulfur to produce their stock-in-trade, phosphors for fluorescent lamps and television screens. Shuji noticed that employees coming in and out of the buildings were wearing white coats stained with red and yellow. Chemical factory equals pollution, he could not help thinking to himself. And with his heart sinking, he wondered: Could I survive at a company like this?

Paying no attention to his student's misgivings, Tada strode off down a small path through the pine forest that led to one of the single-story, galvanized iron-roofed buildings. Inside was a reception area where they were served tea. Then Nichia's president, Nobuo Ogawa, appeared. After a perfunctory introduction—"This is Shuji Nakamura, please be good to him"—Tada immediately struck up an enthusiastic conversation with his old friend. The pair went on and on about their student days, completely ignoring the young man. Understandably tense, Shuji just sat there and listened.

Eventually, it was time for lunch. The three went out for their midday meal a local restaurant. Nakamura could not help but be impressed by such extravagance. An unsophisticated country boy like him was not used to eating out. Over lunch, the two old men—both at the time in their mid-sixties—continued swapping reminiscences, taking no notice whatsoever of Nakamura. After the meal Ogawa said goodbye, and the pair headed back to Tokushima. All the way home Nakamura fretted: Would Nichia really hire someone like him, whose speciality was electronic engineering, a completely different field from chemicals? He felt utterly helpless.

Back at the university, Tada asked him what he was going to do. Nakamura replied that he had already made up his mind. April was almost upon him. On graduation, he had to find a job in order to feed his family. He was in no position to pick and choose: a position at this small local chemical company was all there was. Once again, he asked his professor to do whatever he could to plead his case.

A couple of days later, while finishing up some work in his laboratory at the university, Nakamura got a phone call from Nichia. It was the president himself on the line. Ogawa said, "Tada introduced you, but are you really willing to come and work here?"

"I'm very keen," Nakamura implored. "Please let me join."

Then Ogawa said, "This is not the kind of company that someone like you should join. Nichia's not good enough for you—you should go and work for a better company, like Kyocera, since you have an offer from them. Our company is not in good shape, we may go bankrupt at any time, so think hard."

But Nakamura refused to take no for an answer. "I'm not too good for you," he pleaded. And after putting down the phone, he went to Tada once again to seek the professor's help.

It turned out that Ogawa had rejected Tada's initial overture because Nichia had already hired a half-dozen new employees, all the company could afford that year. But, knowing that Ogawa was always on the lookout for smart workers, Tada tried a different tack, telling him that Nakamura was the top student in the engineering department and a diligent worker to boot. This recommendation must have done the trick. When Nakamura phoned Nichia's president a few days later, to reiterate that he was determined to join the company, Ogawa sighed and said, "In that case, it can't be helped." In due course, a letter arrived confirming Nakamura's appointment.

Little did Shuji know it then, but as subsequent events would prove, joining Nichia was the best career move that he could possibly have made. However, this would not be immediately apparent. In the interim, the young man would have to serve an agonizing ten-year apprenticeship in the research, development, and commercialization of conventional infrared and red light emitting diodes, their component materials, and their intermediate products.

In the previous two decades, much work had already been done on these materials at companies and universities around the world. There was thus little in the way of new advances that Shuji could make. His first few years at Nichia would consist of a series of bitter frustrations until finally, unable to stand it any longer, he would demand to be allowed the freedom to do what he wanted to. When that moment came, remarkably, Ogawa would give Nakamura his blessing.

* * *

Nichia Chemical Industries, the company that Shuji had joined faute de mieux, was founded by Nobuo Ogawa in 1956. The son of dirt-poor rice farmers, Ogawa was born in 1912 in the small Shikoku town of Anan. He managed to escape the drudgery of farming via a military scholarship to Tokushima Technical College, as the university was then known, to study pharmacology. Following further qualification at a military medical college, Ogawa was commissioned as a pharmaceutical officer in the Imperial Japanese Army's ambulance corps, supplying drugs to military doctors. Among other places, Ogawa served at the battle of Guadalcanal. There, as many years later he liked to tell visitors, the forces fighting the Japanese included the future US president, John F. Kennedy.

Even in the army, Ogawa was already demonstrating his creative flair for making things. He contributed significantly to the frontline production of Ringer's solution, which is used to resuscitate bodily fluids after blood loss. In Tokyo he improved the phosphor plates for x-ray equipment used to diagnose tuberculosis. After the war and his release from the army, Ogawa worked briefly for an oil refinery. But his destiny was to be an entrepreneur. In 1948 he returned to Tokushima where he set up his first company, a pharmaceutical laboratory. Ogawa started literally from scratch, without a house, without even a bed, with nothing other than bright ideas and an unquenchable determination.

In 1951 Ogawa came up with an improved process for the production of high-quality anhydrous calcium chloride, a chemical used at that time in the production of the new wonder drug streptomycin. The problem was convincing skeptical drug makers of the superiority of his product. After a great deal of effort and significant personal hardship, Ogawa succeeded in gaining the approval of Japan's leading maker of the antibiotic. But it turned out that the streptomycin market was not large enough, forcing Ogawa to look for other applications for his material. He hit upon calcium phosphate, which is used in the production of phosphors—inorganic luminescent materials that glow when electrically excited—for fluorescent lamps. Ogawa had seen and been impressed by fluorescent lamps for the first time during the war in Mindanao, in the Philippines. He vowed that, when the war was over, he would make something like that.

A fluorescent lamp contains mercury vapor that, when ionized, emits ultraviolet light. This in turn excites a phosphor that emits white light. Tubes coated with the improved phosphor, made using the process Ogawa created, shone with a light that was more than 20 percent brighter than that of conventional fluorescent lamps. He established Nichia Chemical Industries to manufacture and sell this new product. The company's name is compounded from *Nichi*, meaning Japan, and *a*, for Asia. Its motto, "Ever researching for a brighter world," would subsequently prove particularly apt. Ogawa started the company with just twenty-two employees. He hired local folks, whose previous job had been making charcoal, and trained them in how to hold test tubes and perform chemical analysis.

As with his previous venture, Ogawa's biggest problem was convincing potential customers that a hick from the sticks could come up with something that was better than the fancy US technology that big Japanese firms like Toshiba were paying so much to license. Despite its manifest superiority, they simply would not buy his stuff. As so often in the annals of successful Japanese start-ups, it took validation by a foreign customer before domestic buyers would touch the locally made product. In the case of calcium phosphate, Nichia's first customers were the American and Dutch lamp makers Sylvania and Philips.

A hands-on industrial chemist, Ogawa thus had direct personal experience of commercial success achieved with homegrown research. From it he derived his own philosophy regarding innovation and how to go about it. Book learning was no good: country boys like him were more practical. "If you study books, you only believe what is written in them, and you can't go on to the next step. Just reading books and copying what's in them is no good. Read the books, then stop and think, and you'll be able to do something better than what's in the book." Asked how he came up with his own inventions, Ogawa replied, "By thinking hard and working hard. Everybody used to do it by the book, without thinking. But that way, you can't make any improvements."

Throughout his life Ogawa maintained a commitment to improving the lot of the local people of Anan, his birthplace (2004 population: 55,472). He often used to say that it would be much more convenient if his company had been located in Tokyo or Osaka. In Shikoku it was customary for eldest sons to follow in their fathers' footsteps and become farmers. Younger sons usually had to forsake the land for the big city in order to find a job. Thus, by providing employment locally, as more or less the only significant company in the area, Nichia played an important retaining role for the community.

Almost all of Nichia's employees lived near the company, many of them close enough to go home for lunch. (Nakamura was an exception. He commuted from Tokushima, a forty-five-minute drive. "How do you manage to commute from so far away?" his fellow workers would ask him bemusedly.) Most of them were farmers' sons, with the family farm still taking priority over work at the company. When the rice had to be planted, the fields weeded, or the crop harvested, employees would come to work late, or leave early. They didn't have to give notice that they were going to be absent; all they had to say was that they had been working in the fields, and nobody would complain.

Nichia was famous—it had been written up in the national press—as the company with the longest summer vacation in Japan. It lasted twenty days, starting from July 25. Not mentioned in the newspapers, however, was how this practice had originated. Ogawa had been forced to send his workforce on extended vacation one year because he was unable to meet the payroll. During these long summer breaks, Ogawa would head off to the Japanese Alps to indulge his passion for mountain climbing. He claimed that good ideas came to him as he ascended the slopes.

By the time Nakamura joined in 1979, Nichia's staff had grown to around two hundred employees. The company had become a major supplier of phosphors not only for fluorescent lamps but also for televisions. (Red, green, and blue phosphors coated on the inside of the screen light up when hit by electrons fired from guns in the neck of the cathode ray tube.) Japanese TV set manufacturers like Matsushita were Nichia's major customers. Still, it was always a struggle, and the company was forever at the mercy of its big clients. Raising capital to finance new facilities was tough for the privately held firm. Budgets were often stretched thin.

When Nakamura joined Nichia, business was bad. So much so that the previous year the company had had to lay off some staff. Shuji recalled that things were so tight that, if you needed a new pencil, you had to go to your section chief and get his approval to buy one. But in good years, when there was cash to spare, Ogawa would pour most of it into research and development. And the old man was very proud of the fact that, in a country notorious for its government-sponsored R&D, he had never accepted a single yen of taxpayer money.

It is tempting to think that in Nakamura, Ogawa saw a younger version of himself. For a while after Nakamura joined Nichia, the president would drop by once a week to ask how he was doing. Later on, the old man certainly took great pride in his number one researcher's extraordinary achievements. But the gap between them was great, both in terms of status—Ogawa was the founder-president, Nakamura merely an employee—and of age. By the time Nakamura made his bright blue breakthrough, Ogawa was eighty years old. The relationship, Nakamura insisted, was not deep.

Nonetheless, in the heady days following the momentous announcement of the bright blue light emitting diode, it was wonderful to catch the two of them together, as I was fortunate enough to do in May 1995, on my second visit to Nichia. (On my first visit, Ogawa had been off on a trip to the North Korean-Manchurian border, climbing Mount Paektusan, a peak that stands 2,750 meters [9,000 feet] above sea level. "Three thousand meters is about my limit these days," he told me. The only sign of

Ogawa's presence that first time was a photograph that hung in Nichia's spartan reception room. It showed the old fellow in full mountaineering gear, carrying a rucksack and an ice axe, heading off up Mount Kilimanjaro, Africa's highest peak.)

Ogawa struck me as a humble, essentially decent man. His voice was loud, his manner confident, and there was a twinkle in his eye. He had a ready laugh and a fund of funny stories. For example, he told me how, when several senior executives at local banks who had refused to lend Nichia money died of cancer one after another, Ogawa was somehow able to suggest to their superstitious successors that, if they wished to avoid the same fate, they should give Nichia a loan. They did. Later, when he bumped into these bankers, he would cheerfully reassure them, "Don't worry, you'll live long!"

Shuji seemed easy in Ogawa's company. Youth did not defer to age, which in Japan is most unusual. The day I was there the atmosphere became giddy as they chuckled at each other's descriptions of events leading up to the bright blue breakthrough. For Nakamura, the great thing about Ogawa was that the old man had been completely hands-off in his approach to his young employee's research. He had never told Nakamura what to do, just given him the funds and the freedom to get on with it.

We talked about Shuji's work. "We don't know what goes on over there," Ogawa said, indicating Nakamura's laboratory. "All we know is paddy fields and plowing!" At this, Shuji dissolved in a fit of giggles. "So even if we do go and look at what he's doing, we still don't know . . . and he doesn't tell us much, either!" More gales of laughter.

Asked why he gave Nakamura the go-ahead to develop bright blue LEDs, Ogawa replied, "He's good at thinking, so my policy was that if I let him get on with it, he'd probably be able to do it." In Ogawa's view, researchers had to have freedom. "Unless you let them do whatever they like, they can't do research. So if they ask you for some equipment, you have to buy it for them." Even if that equipment was very costly: in Nakamura's case, Ogawa let him have about $2.4 million,* an astonishingly large amount, equivalent to 2 percent of the company's sales that year.

---

*All dollar amounts in US dollars, unless otherwise indicated.

"Of course it was a big risk," Ogawa told me, "but then research is synonymous with risk." As he knew better than most, everyone who ever created anything must be a gambler.

∗   ∗   ∗

Japan is an archipelago, made up of four main islands. Honshu, by far the biggest of the four, is where most of the population lives, crammed into megacities like Tokyo, Osaka, and Nagoya. Hokkaido in the north is where people go for skiing holidays. Kyushu in the south, while styling itself "Silicon Island" on account of the microchip factories located in Kumamoto and Oita Prefectures, is in fact mostly rural. But at least Kyushu has major cities, like Fukuoka and Nagasaki. Shikoku, the smallest and least populous of the four islands, nestles beneath the southern end of Honshu, on the Pacific Ocean side of Japan. Shikoku is thus farthest away from the Asian mainland, making it doubly remote. The construction during the 1980s and 1990s of three enormous (and enormously expensive) bridges linking Shikoku with the mainland was supposed to open up the region to economic development. However, high tolls have kept the volume of traffic across the bridges down, with many drivers still opting to take the slower-but-cheaper ferries.

"Shikoku's high mountains and steep slopes severely limit agriculture, habitation, and communication. . . . Much of the island is a thinly populated agricultural region, with few natural resources and little large-scale industry." The island's climate is humid sub-tropical. A mild winter permits the growing of out-of-season fruit and vegetables, such as tomatoes and cucumbers, for sale to the affluent consumers of nearby Osaka, Kyoto, and Kobe. Such produce is grown in what Japanese call "vinyl houses"—that is, greenhouses in which translucent plastic replaces glass.

I noticed these plastic sheets glinting in the morning sunshine as I flew down to Tokushima for my first meeting with Shuji, in mid-September 1994. From Tokyo, where I lived, there were very few flights to choose from: two going in the morning and two coming back in the afternoon.

In San Francisco that summer I had suggested to John Battelle, my editor at *Wired* magazine, that Shuji's bright blue breakthrough would be worth a feature. John agreed. I faxed Nichia to set up an interview, and was surprised to receive a reply from Shuji himself. This seemed to me charmingly informal. Normally—that is to say, at the big Japanese companies I was used to dealing with—a corporate PR person would have been delegated to handle such arrangements. But perhaps Nichia was not yet big enough to have a PR section? Looking back today, it seems to me that the fact that Nakamura was arranging his own schedule was actually indicative of his isolation within the company.

On that first visit to Nichia, and on a subsequent trip the following year, I made the fifty-minute ride from Tokushima City to Anan by taxi. It was a depressing journey. Though Tokushima is a rural prefecture, the road to Anan passes through the kind of rundown, decidedly unpicturesque, semi-urban sprawl that one so often encounters in the Japanese hinterland. Interspersed between the rice paddies and the vinyl-covered vegetable patches was the usual jumble of gaudy pachinko parlors, fast-food outlets, and used-car lots. I remember noting along the roadside one particularly sad and dilapidated-looking coffee-shop-cum-karaoke-bar that had been built in a crude approximation of a space shuttle.

On my third trip four years later, however, Shuji sent some of his juniors in a company minivan to pick me up at the airport. They drove me to Nichia via narrow back roads. In the bright midmorning sun, the Shikoku countryside looked fresh and beautiful, with verdant fields in the foreground, picturesque limestone mountains looming behind. It must have rained recently because I distinctly recall fording a rushing stream. This was more like it: now I began to understand why, twenty years before, the nature-loving Nakamura had been so reluctant to forsake his native island for the overcrowded concrete jungles of the Japanese mainland.

Arriving at Nichia on that first occasion, I was met in the reception area by Shuji himself. Again, this was unusual: ordinarily you would be shepherded into a meeting room by some corporate functionary, who would then summon the interviewee. The functionary would remain for the duration of the interview, making sure that nothing untoward was mentioned, that the company line was adhered to. At Nichia, however, Shuji came alone and we spent an hour or so together, just the two of us chatting away.

Nakamura turned out to be a man of medium height and spare build dressed in the gray battledress fatigues of corporate Japan. His mien was youthful, his features agreeable, with a high forehead topped by a full mane of unruly black hair. Though he had just turned forty, his hair had not yet begun to turn gray, but it was beginning to recede. What struck me most about Shuji on that initial encounter, however, was not his appearance, but his disposition.

Your stereotypical Japanese salaryman is shy, inarticulate, and unused to speaking out. He is by nature suspicious and somewhat secretive. His manner is formal and he is uncomfortable with strangers, especially foreigners, at ease only in the clannish company of his peers. He is highly conscious of the corporate pecking order and of his place within it. Nothing if not conventional in his tastes, he is reluctant to make decisions, preferring to defer to the consensus in order to maintain group harmony. And if he does have a sense of humor, your average Japanese *apparatchik* is certainly not about to reveal it in public.

Shuji is, by contrast, cheerful and friendly. He is open, frank, and wonderfully outspoken. Ask him a direct question and he will unhesitatingly give you an honest answer. He speaks fast, his voice accelerating in pace and rising in pitch as he becomes excited, a not-uncommon occurrence. By nature a sociable person, he is happy to chat casually with almost anyone, regardless of how long he has known them or their station in life. He is not, as he is the first to admit, sophisticated or streetwise, a certain naivete adding to his charm. On occasion he can also be blunt to the point of abruptness. Despite his extraordinary achievements, he comes across as a modest man, not at all arrogant. Unlike most Japanese, who prefer to be known formally by their family name, pretty much everybody who knows him calls Shuji by his first name (which Americans, perhaps unaccustomed to dealing with diphthongs, characteristically pronounce as "Sooji"). He makes his own decisions, is beyond question his own man.

Nakamura has been accused by his detractors of being "un-Japanese." Not being Japanese myself I cannot judge whether this accusation has merit. Certainly Shuji is an individual who thinks for himself in a society where conformity and unquestioning obedience are the norm.

But can it really be un-Japanese to have a distinct personality, strong opinions, and the courage to voice them?

Though essentially a serious person, Shuji also sees the funny side of life. He has a broad, slightly goofy smile and he loves to laugh, especially at his own expense. A conversation with Shuji is invariably punctuated with his distinctive loud, high-pitched, panting laugh. At points in the narrative that strike him as particularly amusing, he will slap his knee and rock back and forth with mirth. My friend and fellow journalist Denis Normile well remembered the first time he encountered Shuji at Nichia. "I have never before or since laughed so hard or so continuously during an interview," he told me.

What was not clear to me from my initial encounter was the degree to which Nakamura was isolated within Nichia. We have already seen that he was unique in his professional speciality, the lone electronic engineer at a chemical company. Also, in that having no assistants, he was responsible for organizing his own schedule. He was, as the Japanese say, *nakama hazure*, which means "separated from the group." Which is not to say that he is a loner. On the contrary, his disregard for seniority means it is easy for him to get along in a friendly way with everyone, especially younger people.

In addition to these professional and administrative matters, there was also a personal dimension to his detachment. Knowing that Shuji was from Shikoku, I had assumed that he was a local Tokushima lad. In fact, he was from Ehime Prefecture, at the opposite end of the island.

✳   ✳   ✳

Shuji Nakamura was born on May 22, 1954, in Oku, a tiny fishing village located two-thirds of the way down the Sadamisaki Peninsula on its Pacific side. A spit of land some twenty miles long, the peninsula is so narrow that in some places you can actually see the Pacific Ocean on one side and the Seto Inland Sea on the other. It resembles a little tail jutting out of the island's rear end. Sadamisaki is remote, even by Shikoku standards. The local climate is mild. This is ideal for the cultivation of citrus fruit, in particular the mandarin oranges for which Ehime is famous throughout the country. Ehime has long been one of the poorest parts of Japan. In 2000, it ranked fortieth out of forty-seven prefectures in terms of household income.

Apparently unchanged for centuries, Oku was a typical village. Fishing was the main occupation, followed by farming. The locals grow yams on steps cut into steep hillsides. Shuji's maternal grandparents owned such a farm. At harvest time, the village kids would go there to help dig up the yams. His grandmother would steam sweet potatoes for the youngsters to snack on. He and his friends were allowed to run wild. They would play in the mountains all day doing the kinds of things that little boys do, like catching dragonflies and pulling off their wings. On the seashore little Nakamura would sit for hours on end, watching the ships as they sailed past.

When Shuji was born, the roads were not yet wide enough for buses. To get to Yawatahama, the nearest town, which was located at the top of the peninsula, the villagers relied on a tiny ferry. If the sea was stormy, the ferry did not run, and the village was cut off. Shuji remembered a pregnant woman having to be rowed in a boat to the nearest midwife. Oku may have been inconvenient, but the village was an idyllic place to grow up, surrounded by the bounty of nature. Shuji would fish from a small pier, using shellfish as bait. After a typhoon, the whole family would go down to the beach, collecting all sorts of marine creatures that had been washed up by the storm.

His father, Tomokichi, worked as a maintenance man for Shikoku Electric Power. The family—Shuji has two older siblings, a brother and a sister, and a younger brother—lived in a company house located next to the transformer substation. Whenever there was a typhoon, his father would have to venture out in the rain to make sure the equipment was undamaged. From his father, Shuji learned how to make wooden toys, like catapults and bamboo propellers. He liked making things, and became good at it, a skill that would later stand him in good stead.

The substation was located at one end of the village, the primary school at the other. For his first year Shuji walked along the strand to school. His teacher was Miss Kono, a young woman fresh out of training college. In his second year at primary school, the idyll came to an abrupt end. The substation was to be automated; no more need for a maintenance man. The family relocated to Ozu, a small inland town about twenty miles away. They loaded their household goods and chattels onto a boat for transfer to a ship anchored offshore. At the pier Miss Kono and all Shuji's classmates came to see him off. The teacher gave him a notebook and some pencils. "Good luck, little Nakamura," she said tearfully. The farewell was, as Shuji recalled decades later, like a scene from a movie. Afterward, he and his siblings would all look back nostalgically on their early years in Oku.

*   *   *

At school Nakamura was by no means an academically gifted student. His boyhood was in fact pretty typical. He fought continuously with his elder brother, Yasunori. Every day the pair had wrestling matches. Smaller than his sibling, Shuji always lost these fights. But though physically defeated, mentally he would never give in. The fights would start over trivial matters, who got what sweets, which TV channel to watch, pretty much anything. Mealtimes were especially contentious. "You got more than I did!" one brother would shout, and a fight would begin. Eventually, in exasperation, their father would throw the pair out of the house. They had to sit outside the door until late at night when their mother would let them back in.

After the family moved to Ozu, Shuji would run around with the other kids, exploring, playing kick-the-can and hide-and-seek, coming home in the evenings all dirty. He loved comic books, his favorite being *Mighty Atom*. His mother was always chiding her boys to do their homework. "Don't watch television," she would shout. "Study!" But for the most part, they ignored her admonishments.

All through primary and high school in Ozu, Shuji's passion was volleyball. At junior high, his elder brother was captain of the school volleyball team, so Nakamura was more or less forced to join. There was no gym at the school, so they practiced in the mud, sliding on their knees to prevent the ball from hitting the ground. Nakamura remembers that his knees bled every day. Lacking a coach, the team did not know much about technique and strategy. They rarely won and were always ranked bottom of the league. Nonetheless, they tried hard, training until dark every night, practicing at school on Sundays. Nakamura thought that if he disciplined his body he would get stronger, but the team kept losing. Fiercely competitive from an early age, Shuji always hated to lose.

Volleyball was so all consuming that it left Nakamura little time to study for his high school entrance exams. He was always bad at rote learning, never managing to memorize names, dates, and places. In an education system heavily geared toward the memorization of information, this was a significant disadvantage. Nonetheless, he was good at math and science. Somehow, he managed to scrape into Ozu High, an academically oriented school whose priority was getting its students into universities. Here, too, however, volleyball remained his priority. He read books on the game, considered different formations and strategies. Still, his team kept losing.

In his second year at Ozu, his classroom teacher told him that if he wanted to improve his scores, he had better quit playing. It was time to concentrate on studying for the all-important university entrance exams. This was good advice. Only dropouts and vocational studies students played sports. Unwillingly, Nakamura gave up his beloved volleyball. But his departure meant that the team no longer had enough players to continue. His teammates begged him to come back. So after two weeks he rejoined the team and resumed full-time practice. He was the only student in the A-stream to continue playing sports until the end of high school.

Nakamura paid a price for his dedication to volleyball. He studied as hard as he could, but his university entrance exam results were not good enough to win him a place at a prestigious school like Tokyo University. Nakamura had always loved theory and logic. His dream had been to become a theoretical physicist or a mathematician. But his teacher told him, You can't make a living from physics—you'd better choose a course like engineering so that you can find a job when you graduate. Once you're at university, you'll be able to study whatever subjects you like. But to Shuji's great disappointment, this turned out not to be the case.

A friend of his was planning to go to Tokushima University. This was a state school, and as it happened, one of the two schools whose entrance exams Nakamura had taken. His friend invited Shuji to join him. Electrical engineering seemed close to physics, so that is what—reluctantly—Shuji chose to study. Though nominally a university, Tokushima was actually little more than a souped-up technical college. When the nineteen-year-old Nakamura arrived there in 1975, many of the professors were former high school teachers. They could barely understand semiconductors, let alone quantum physics. Nor were the facilities any better. For study the students only had tatty, out-of-date textbooks.

The first two years at Tokushima consisted of general studies. These included arts courses, which Nakamura hated. He couldn't understand why he had to take such irrelevant subjects. After about a month, Shuji stopped going to classes. He gave up part-time tutoring because it was a nuisance. He cut himself off from company, telling three close friends from high school who had entered Tokushima the same year that they were no longer welcome to come and visit him. "Don't drop by anymore," he said. "You're distracting me."

After that, Shuji withdrew to his room. All day long he would read books, mostly on physics, more or less in solitude. For him, it was a time for reflection. He had always liked thinking. Now, in between reading, he thought, as young students do, about matters philosophical. Reading and pondering what he had read, he was learning in his own way. But there was a limit to the amount he could learn by reading alone.

Though bitterly disappointed at the quality and the content of his courses at Tokushima, Shuji realized that there was no use crying over spilt milk. His parents were paying his fees, his mother working part-time to help support him. This withdrawn solitary way of life could not continue indefinitely. "I decided that I would graduate as quickly as possible from this crap school so that I could support myself and be independent." He started going to lectures again. He studied hard, attended classes properly, and managed somehow to scrape through the subjects he didn't like.

Finally, in his third year at Tokushima, Shuji found something that interested him. He attended a lecture on semiconductors given by an assistant professor in Osamu Tada's laboratory. Insecure about venturing out into the real world and fascinated by the physics of solid-state materials, he decided to stay on at university for a further two years and earn a master's degree. He asked Professor Tada to be his thesis adviser. Until then, he had only been listening to lectures, doing paper calculations. Now, for the first time, Shuji started doing experiments on his own. Things were happening in front of him that he could measure and evaluate. Real learning at last!

As his thesis topic, Nakamura chose the conductivity mechanism of the compound barium titanium oxide. His focus was resolutely theoretical. In attempting to clarify his results from a theoretical point of view, he loved reading technical papers and poring over reference materials. But Tada was a dyed-in-the-wool experimentalist. He would catch his student reading papers and tell him that it was no use being armed with theory if you couldn't make actual things. "Papers are no use once you're out in the real world," Tada would insist. No wonder he and Ogawa got on so well.

At the university, Tada's lab was known as "the junk room." It was filled with piles of unwanted electronic equipment—broken-down television sets, old radios, anything that could be cannibalized for spare parts. On one occasion, Nakamura asked his prof to buy him an electronic circuit he needed. Pointing to a heap of trash, Tada replied, "There's the part you want." Taken aback, Nakamura asked, "Where?" And Tada reached into the pile, pulled out a broken radio, and cried triumphantly, "Here!"

The lab had no budget to purchase proper equipment. As a result, in order to build the things they needed, students were forced to acquire all sorts of manual skills—soldering, cutting and joining glass, beating and welding sheet metal, fashioning parts on a lathe. Tada's students may have had to get their hands dirty, but under his guidance they became very resourceful. Nakamura remembers his years as a graduate student as akin to being a sheet metal worker in a small factory. What he really wanted to study was theory. But most days much of his time was taken up with jury-rigging equipment for experiments. Sometimes, as he was reluctantly grinding a part, getting oil all over his fingers, he would curse under his breath: "Shit! Why do I have to do this sort of thing?"

In fact, Nakamura was gaining precisely the kind of self-reliance and skills that he would need in his quest to develop a bright blue light emitting diode. At Nichia, he would be forced to make or modify much of his own equipment. Ultimately it would largely be this technical mastery that would give him the edge on his rivals.

Otherwise, his university days passed unremarkably. He boarded in the house of an old high school friend's family, his only possessions a heater and a radio. He had no money for after-hours entertainment, and no friends. All he did was sit by himself in his room and study. For recreation, Nakamura would go jogging along the banks of a nearby river, running between six and twelve miles a day. Then he would go to a public bath, eat his evening meal, and cogitate.

One evening in his third year, Shuji was in the university canteen wolfing down his usual dinner—the hundred-yen set meal that was the cheapest item on the menu—when he heard music coming from a small hall next door. Curious, he stuck his head round the corner. He noticed a pretty girl standing by herself and, surprising himself with his own audacity, asked her to dance. Her name was Hiroko and she was an education student, a banker's daughter, also in her third year. They got on well, and one thing led to another. The following year, Hiroko graduated and began work as a teacher at the university kindergarten. Then she became pregnant.

Hiroko and Shuji were married on February 22, 1978. A student wedding was so unusual that it made the local newspaper. Since Nakamura was still a penniless student, the reception was a very humble affair, with Professor Tada serving as the master of ceremonies and each guest contributing a measly two thousand yen (about five dollars). Given their lack of cash and Hiroko's condition, there was no question of a honeymoon. Hitomi—the first of three daughters—was born in August. After maternity leave, Hiroko resumed work, leaving the baby in the care of her mother. Shuji moved in with his in-laws. Then, as we have seen, he started looking for a job, which would ultimately lead him to Nichia.

There, over the next decade, he would be assigned various tasks, all of them related to light emitting diodes. In each he would succeed spectacularly well from a technical point of view, only to fail disastrously from a commercial one. But in order for us to understand what it was that Shuji was trying to achieve, we must first look at the LED itself, what it is, how it works, and its origins and early history.

CHAPTER 2

Arthur C. Clarke once posited, "Any sufficiently advanced technology is indistinguishable from magic." I think I know what he meant. As a nonscientist, a completely untechnical person, I have always regarded semiconductor devices—microchips if you will—as magical. After all they are just tiny pieces of . . . *stuff*; all solid-state, no moving parts. Yet pump an electric current through them, and they . . . *do* things. Like amplify sound, or memorize information. And, for as long as I can remember, semiconductor devices have kept getting better. Or, in the case of light emitting diodes, brighter.

With LEDs, I have similar feelings, only more so. Hook a light emitting diode up to a battery and you can actually *see* what it does. I was delighted to discover that LED researchers themselves are not immune to this sense of wonder. For example, John Kaeding, one of Shuji Nakamura's postgrads, is showing me around Shuji's basement lab at UC Santa Barbara. Under a microscope, Kaeding touches two probes to a tiny dot on a sliver of wafer, producing a bright green glow. "It's a magical moment," he muses. "We know how these things work but still, there's something about it that seems magic."

"The best thing about LEDs, you can see the result, and that's the magic," Warren Weeks, an ace crystal grower who formerly worked for the US blue LED specialist Cree, tells me as we sit sipping drinks on the roof of a hotel in his hometown of Charleston, South Carolina. "I've also worked on radio-frequency devices and they're not nearly so much fun—you've got to hook them up to an oscilloscope to find out what they're doing. With an LED, you apply the voltage and you can really *see* the brightness."

In the past, in order to produce colored lights like the red-amber-green ones in traffic signals, you had to stick a filter in front of a white incandescent lightbulb. Filtering is cheap but inefficient, losing up to 80 percent of the light in the process. LEDs by contrast produce light that is colored to begin with, spiking at a specific wavelength. Unlike the often insipid-looking colors produced by filtered incandescents, especially old ones, LEDs shine with a pure, "saturated" hue (i.e., one not diluted with white).

On my desk in front of me as I write is a bright blue LED that Shuji Nakamura gave me on my second visit to Nichia in 1995. I have it connected, via a resistor, to a 9-volt battery. When I examine the device closely, most of what I see is package, not diode. The package is a solid transparent bell-shaped epoxy sheath about the size of the eraser on the end of a pencil. It serves to encapsulate the device and, through its dome lens, to disperse the light. Inside the plastic casing, about halfway down, are two vertical legs sculpted from silvery metal. These are terminals that connect the LED to its power supply. One—known as the anvil, due to its appearance—contains a recess into which the LED chip is glued. The recess is cup-shaped so as to throw the light upward. The other terminal is known as the post. It is just possible to perceive the two hair-thin wires that connect the chip to its terminals. The chip itself is almost impossible to see. When lit, it is too bright; when not lit, too small. Most LED chips are less than a millimeter square, about the size of a grain of sand.

When Asif Khan was showing me around his impressive lab at the University of South Carolina, I asked to see what gallium nitride—the basic material from which bright blue, green, white, and ultraviolet LEDs are fabricated—looks like. Flipping open a little plastic box, he used tweezers to pick up a round wafer. It was about the diameter of an Oreo cookie, but much thinner and with one edge lopped off. The wafer was made of sapphire. We associate sapphire with blue, but that color comes from naturally occurring impurities in the crystal. Man-made sapphire wafers such as the one Asif showed me contain no impurities, and hence are completely transparent. When he tilted the wafer to catch the light, iridescent patterns of pink and green became faintly visible. These indicated the presence of thin films of gallium nitride and related compounds. Invisible devices grown on a transparent substrate? Magic, to be sure.

The whole thing seems implausibly delicate, most unlike the schematics that I am familiar with from leafing through the technical literature. These typically depict LEDs as layer upon layer of materials, each of a slightly different chemical composition, the whole somewhat resembling a club sandwich. It turns out that many of the layers are only a few atoms thick, hence their apparent insubstantiality. Only when the wafer is etched so that metal electrodes—most commonly thin films of gold—can be sputtered (sprayed) on does it lose this aura of invisibility. On two-inch wafers, the current industry standard, you can fabricate approximately twenty thousand LED chips; on three-inch wafers, to which the industry is now moving, about forty-five thousand. There is still considerable variance in device performance across the wafer, hence there is plenty of room for LED manufacturers to make cost-reducing improvements to the production process for many years to come.

Via the electrodes, every chip is tested using a wafer probe. An X-Y grid map is generated, then the wafer goes to a breaker, or a laser separator, to be sawed up into individual chips. The map is fed to a chip sorter, which sorts the chips at high speed into different bins by brightness, wavelength, or voltage. This process is in itself mind-boggling—how *do* you sort grains of sand?—but it need not concern us here. Suffice it to say that the best devices go to high-end applications like automotive headlamps or computer backlights, while the off-spec stuff ends up in cheapo-cheapo applications like decorative jewelry and Christmas tree lights. LEDs are packaged in various ways by specialist firms, many of them located in China.

*   *   *

What exactly are LEDs, these tiny specks of magic material, and how do they work?

A modern light emitting diode is simultaneously one of the simplest and the most sophisticated of electronic devices: simple in the sense that it is a diode, a device that by definition conducts electricity in one direction only. This property makes diodes useful for converting alternating current to direct current, a process known as rectification. Unlike a transistor, which has three terminals, a diode has but two.

In its most basic form, a light emitting diode consists of two layers made of the same material. One is negatively charged, that is, doped with a tiny dose of impurities to give it an excess of electrons; the other, positively charged, doped to give it an excess of what electrical engineers are pleased to call "holes."

Holes are represented schematically as particles, but in reality they are not. In fact, holes are merely the absence of electrons. Gallium nitride pioneer Herb Maruska suggests a helpful analogy. Think of an atom as an egg carton, with some eggs missing. The eggs are electrons; the sites where there are no eggs, holes. A single egg can move much faster around eleven empty spaces than a hole surrounded by eleven eggs. This explains why electrons are always much more mobile than holes.

On the application of a voltage, excess electrons flow across the junction between the layers from the negative side of the LED. Once they reach the positive side, some of the electrons combine with some of the holes. They annihilate one another, giving up the ghost in the form of a photon, i.e., light.

The sophistication of an LED lies in the fact that, in order to get the devices to emit bright light, you have to optimize them. Instead of a single positive-negative junction, you have to build what is known in the jargon as a "double heterojunction" (aka "heterostructure"). This ponderous locution simply means a sandwich consisting of two layers of a slightly different material that are used to confine the active, or light emitting, layer. This layer, in its thinnest form, is known as a *quantum well*.

Early LEDs could only produce one photon for every thousand electrons. The point of quantum wells is to improve the chances of electrons bumping into holes. The wells are extremely thin, just a few atoms deep. These tiny, almost two-dimensional trenches serve to trap the mobile charge carriers. Unable to escape from the trench, particles and absences of particles are forced to combine. If you like, you can imagine electrons as drops of liquid falling down the well and holes as bubbles rising up it.

(What makes a well quantum? As with anything to do with quantum mechanics, if like me you are untechnical, this is a question that it is probably better not to ask. In essence, "quantum" means something that is very small indeed.)

A light emitting (or laser) diode may contain any number of quantum wells, but the optimum number for mopping up all the mobile charge carriers in an LED seems to be three. This means that modern devices consist of at least ten layers: quite a sandwich. To grow quantum wells of atomic-level thinness and with sufficiently abrupt transitions between one layer and the next requires, as we shall see in the next chapter, extremely sophisticated computer-controlled equipment.

It helps, from the crystal grower's point of view, if you are good at visualization. "To be a good materials scientist, you have to be able to see the atom, you have to kind of *be* the atom," explains Warren Weeks. "You have to say, Hey—what's this temperature going to do to a little atom on the surface of the wafer? And you get a feel for how chemical bonds break apart, for temperatures and gas flows, you can kind of visualize the atoms in the gas phase, because the reaction is going from gas to solid, so it's kind of strange."

Light emitting (and laser) diodes are made from materials called semiconductors. The name comes from the fact that semiconductors sit midway in the spectrum of materials between conductors, like metal, and insulators, like glass. The best-known and most common semiconductor is silicon. It has four bonding electrons in its outermost shell, known as the conduction band, unlike conductors, which have eight, and insulators, which have none.

In addition to *elemental semiconductors* like silicon, it is also possible to synthesize *compound semiconductors* out of two elements, one having three free electrons, the other five, like gallium arsenide, gallium phosphide, or gallium nitride. The one with five (arsenic, phosphorus, nitrogen) donates an electron to the one with three (gallium), so that both have four, just like the elemental semiconductors. In addition to these so-called three-five compounds, you can also synthesize two-six compounds like zinc selenide.

To give an excess of electrons or of positively charged holes, to make the material negative- or positive-type, trace amounts of impurities known as *dopants* are added to the mix. In gallium nitride devices, silane—a derivative of silicon, which has four bonding electrons—is typically used to make n-type material (a semiconductor with extra electrons), while magnesium, which has two bonding electrons, supplies the holes.

All light emitting (and laser) diodes are made from compound semi-conductors. Why go to the trouble of using a complicated material when a simpler, much cheaper one like silicon is available? The answer is that silicon does not normally emit light. It suffers from what is known as an *indirect bandgap*.

The bandgap is the amount of energy it takes to jolt an electron up from the valence band (where it is bound) to the conduction band (where it is free to move, and hence combine with a hole). The bandgap also determines the energy of the photon produced by the combination, hence the wavelength of the light emitted.

The higher the energy, the shorter the wavelength. Gallium arsenide, which emits infrared light at around 885 nanometers, is a narrow-bandgap material; gallium nitride, which on its own emits ultraviolet light at around 365 nanometers, is a wide-bandgap material. Nanometer wavelengths, produced by quantum wells just a few atoms thick? That explains why it is sometimes asserted that LEDs are the world's first ubiquitous nanotechnology. The degree of purity the materials have to exhibit is staggeringly high. Impurities on the order of two or three parts per million are enough to stop the show, as are nanoscale cracks in the crystal lattice.

In *direct-bandgap semiconductors* like gallium arsenide and gallium nitride, electrons zip across the gap between valence and conduction bands and combine with holes unobstructed. In an *indirect-bandgap semiconductor* like silicon carbide, the interaction is much more complicated. So much so that it requires the crystal lattice to shudder. Combinations still take place, but most do not produce light. Indeed, it is hundreds of times easier for light emitting combinations to occur in a direct-bandgap semiconductor than in an indirect one. Hence, as we shall see in the next chapter, Shuji Nakamura's decision to eschew silicon carbide in favor of gallium nitride. Though a blue light emitter, SiC has an indirect bandgap, thus by definition cannot produce bright light.

So much for colored light emitters. But how do you get white light, which by definition is colorless, out of colored LEDs? The answer is, you can do this in two ways by adding lights of different colors. The first way involves mixing light from red, green, and blue LEDs (amber is sometimes also added). This delicate balancing act is relatively expensive. The second way is simpler and much cheaper. You take complimentary colors, a blue light emitting diode and a yellow (yttrium aluminum garnet, or YAG) phosphor coated on the inside of the LED's plastic casing. The YAG phosphor converts the light from the LED. The resulting blend is perceived as a cool white. Adding a second phosphor that emits in the red region produces a warmer white. Most of today's white LEDs are made this way. However, the efficiency of warm white devices is often much lower than that of cool white ones. Much research today goes into developing phosphors with better light conversion efficiency.

✳   ✳   ✳

Man's fascination with luminescence dates back to ancient times. Aristotle and Pliny the Elder recorded light emitted by fungus and by the scales of decaying fish. Electroluminescence was first observed in 1907 by Henry Round of New York. He noticed a "curious phenomenon," namely, that by applying a voltage to crystals of sandpaper grit—carborundum, aka silicon carbide—he could produce a yellowish light. The first report of blue electroluminescence, also in silicon carbide, dates back to 1923. Such naturally occurring light emitters were the direct ancestors of the dim blue silicon carbide LEDs made by Nichia's American archrival, Cree.

In the mid-1950s, at the dawn of the semiconductor era, RCA's Rubin

Braunstein observed infrared light emission from the then brand-new compound semiconductor, gallium arsenide. But the first person to fabricate a visible-spectrum LED and, moreover, to recognize the significance of what he had achieved in terms of its momentous potential applications, was a feisty thirty-four-year-old electrical engineer from the coalfields of southern Illinois named Nick Holonyak.

Holonyak made this first visible red LED in the fall of 1962, with funding from the US Air Force. He was then working at General Electric's Solid-State Device Research Laboratory in Syracuse, New York. In the February 1963 issue of *Reader's Digest*, Holonyak told an interviewer, "We believe there is a strong possibility of developing [the LED] as a practical white source." The article went on to predict, "If these plans work out, the lamp of the future may be a piece of metal the size of a pencil point, which will be practically indestructible, will never burn out, and will convert at least ten times as much current into light as does today's bulb." Prescient words.

Ironically, Holonyak would shortly leave GE, the company Edison founded, after his boss more or less showed him the door. Holonyak had been fooling around in domains that were of no interest to GE management. He returned to the University of Illinois, where he remains, still active to this day. Thus, at a time when GE was way ahead of everybody else, the company turned its back on LEDs, the technology that would eventually displace the lightbulb, GE's original raison d'être. "You can look at that and say, Are those people crazy that they didn't see that the semiconductor had a future in light emission?" an exasperated Holonyak would ask me thirty years after leaving GE. "Why didn't GE pursue it? I mean, they gotta be nuts, when you think about it." But building businesses from scratch is evidently not GE's forte.

The invention of the light emitting diode was a by-product of a race to see whether, in the wake of the recent invention of the ruby laser, it would be possible to produce a semiconductor equivalent. Four groups participated in the race, including one from IBM and Holonyak's. It was ultimately won by Holonyak's colleague and friend at GE, Bob Hall. Having won, however, Hall promptly lost interest in semiconductor lasers, and went off to do something else.

The race to build the semiconductor laser was not motivated by any idea of what such a device would be useful for. The first laser diodes were mere laboratory curiosities, capable of operating only under highly restricted conditions. These included being dunked in liquid nitrogen and zapped with short pulses of current. Indeed, for many years the laser would be dismissed as "a solution seeking a problem." Not until the 1980s, with the appearance of CD players and fiber-optic communications, would the tiny devices finally come into their own.

Early laser research was curiosity driven, in an era when big corporations like AT&T, GE, IBM, RCA, and Westinghouse were able to recruit the cream of the scientific crop. They lured scientists and engineers, not with stock options—an incentive that in 1962 had yet to be thought up—but with hefty salaries and freedom to pursue their instincts, wherever they led, unconstrained by commercial pressures. For researchers in the 1960s, the central labs of corporations were much like the universities where they would normally have worked. Corporate scientists published papers, attended conferences, even took sabbatical years just like professors. Today, when research horizons in the corporate sector are mostly measured in months, and blue-sky research is once again the exclusive domain of academe, such freedom seems very old-fashioned. Back then, however, it was the norm.

Holonyak was more down-to-earth than many of his peers. Unlike Bob Hall at GE, he would never drop research on light emitting devices but would continue to work on them throughout his long career. Holonyak had grown up poor, the son of immigrants from Eastern Europe, from what today is western Ukraine. His father was a coal miner. When little Nick was five, his dad gave him a pocketknife. With it the youngster learned how to make things for himself—slingshots, scooters, whatever: "You didn't go ask for things, you made them."

In 1953, when transistor pioneer John Bardeen invited Holonyak to become his first postdoc at the University of Illinois in the then brand-new field of semiconductor research, Nick jumped at the chance. "We started in a bare room. We had to build everything: benches, all of our equipment. . . . And [we] learned everything from scratch." This ability to build anything would stand Holonyak in good stead when it came to fabricating devices. Much as, almost forty years later, Shuji Nakamura's skill with his hands would help him to scoop the world with his bright blue breakthrough.

The first LEDs were made of gallium arsenide, which as we have seen emits infared light. That was good enough for his peers. But Holonyak insisted that, as he put it, "Our concentration should be on the visible spectrum, because that's where the human eye sees." Accordingly, in order to widen the material's bandgap, Holonyak added a dollop of phosphorus to the mix. He chose an unconventional growth method known as vapor phase epitaxy. "People told me that, had I been a chemist instead of an electrical engineer, I would have known that growing a crystal this way was impossible," he would recall many years later. "No one in their right mind would have tried it." Nonetheless, Holonyak's unorthodox method worked. The result was gallium arsenide phosphide, or GaAsP, the world's first-ever semiconductor alloy. It emitted a dull red light.

GE offered these first visible diodes for sale, via the Allied Radio component catalog, at $260 each. Lasers cost $2,600, an indication of the perceived difference in value between the two types of diode. For many years, LEDs were seen by researchers as the poor relation. Lasers, with their more complex structures (needed to amplify the light) and well-defined beams, seemed much sexier. Today, however, in terms of their applications and production volumes, LEDs are by far the more important device.

Holonyak's main contribution to the field would not be commercial products, but outstanding PhD graduates. At last count, he had trained over sixty doctoral students. First among them was George Craford, a self-effacing country boy from the corn state of Iowa who found physics more to his liking than the arduous family business of farming. He encountered Holonyak during a tour of Holonyak's lab at Illinois. "I saw him stick this little LED, a red speck of light that you could hardly see, into a Dewar flask of liquid nitrogen and, suddenly, there was a bright light that made the whole Dewar glow." Beguiled by this magical demonstration, Craford signed up with Holonyak as his thesis adviser.

Over the years, professor and PhD would remain in close contact, speculating about the future of LEDs and how their devices would eventually replace Edison's lightbulb. "Nick and I have been talking about this stuff for twenty-five, thirty years," Craford told me in 1994. "We've dreamed about it happening; the whole time we've been wondering—if you can replace tubes with transistors, why not tungsten lamps with LEDs? That thought has been there more or less since the beginning, although it seemed fairly preposterous early on."

The first firm to commercialize LEDs, using technology transferred from Holonyak's lab, was the chemical company Monsanto. "They were in the fertilizer business; they basically had a phosphorus mine," Craford recalled. "And they thought, Gee, there must be a use for phosphorus in all this emerging electronics stuff. That's how they got into making compound semiconductors, which is a pretty bizarre chain of events."

Craford joined Monsanto in 1967. Two years later, he achieved a breakthrough, doping a device with nitrogen to produce the world's first yellow LED. "The red LEDs we were selling at that time, you basically wanted to be in a dark room to get the full effect. But this thing, boy, it really lit up in that dark room. It was a yellow device and it was vastly brighter than anything we had seen before." The fact that the human eye is more sensitive to yellow than to red light helped. But Monsanto's marketing people were not impressed. "They came back with the feedback, Gee, our customers are using red, they like red, and they don't much care if it gets brighter, unless it's cheaper. If you can make it cheaper, that's good, but any other stuff is pie-in-the-sky."

The first applications for red LEDs were as on/off indicators and seven-segment alphanumeric displays in calculators and digital watches. As a leading maker of calculators and other scientific instruments, Hewlett-Packard took great interest in the new technology, hiring Holonyak as a consultant and initiating a collaboration with Monsanto.

The third major US manufacturer of first-generation LEDs was AT&T. The phone company needed miniature, long-lasting, low-power indicator lamps for its multi-line telephones. On such phones, the lamps show which lines are being called and which are busy. A second application was to light up the keypad, so that phones could be used in dark places such as bedrooms.

Bell Laboratories developed its own technology, based on the indirect-bandgap compound, gallium phosphide. In addition to red, GaP LEDs could also be made, with the addition of a small amount of nitrogen, to emit a yellowish-green light. Some forty years on from their commercial debut, green gallium phosphide and red gallium arsenide phosphide LEDs are still manufactured and sold in huge quantities, for example, as the digital displays on clock radios.

For battery-powered calculators and watches, however, LEDs were simply too dim to see in daylight and used too much power. By the end of the 1970s, they had been replaced by liquid crystal displays. In telephone keypads, however, LEDs continue to be used, the difference being that these days, they are mostly blue or white, not green or red, and made from gallium nitride.

A new market like LEDs attracts hungry firms. Big Japanese manufacturers including Matsushita, Sharp, and Toshiba soon piled in. A vicious price war ensued. When the smoke cleared, the Americans—notably Monsanto and Texas Instruments—had quit the field. The last remaining US LED standard bearer was Hewlett-Packard, which George Craford joined in 1979. HP begat Agilent, which in 1999 formed LumiLEDs Lighting, a 50/50 joint venture with Philips of the Netherlands. In 2005 Philips bought out its American partner, and LumiLEDs became a wholly owned subsidiary of the Dutch firm. In the interim, however, a feisty new US challenger had sprung forth, in the shape of a start-up called Cree Research, based in North Carolina. Its specialty was blue LEDs. This firm has an important role to play in the saga of Shuji Nakamura and the solid-state lighting revolution. We shall learn more about Cree in chapter 6.

Among Japanese LED manufacturers there was also an unfamiliar name. Stanley Electric was a specialist maker of lamps for car companies like Toyota. Stanley's director of R&D, Toru Teshima, had long been haunted by the notion that a new and better form of lamp than the incandescent lightbulb might come along and destroy the company's core business. The appearance of the LED confirmed Teshima's worst fears. At his urging, the company threw itself into developing the new technology. By 1982 Stanley was producing the brightest red LEDs that anyone—

including Craford, who collected a couple of them from Teshima on a visit to Tokyo—had ever seen. They were so bright it hurt your eyes to look at them, the first high-brightness LEDs.

Stanley's primary target was the car market. This had long been the goal of LED makers. Monsanto had been so optimistic about the future of LEDs in automotive applications that in 1973 the company placed an advertisement in the *Wall Street Journal* showing a car with LED head-lights. In fact, it would be twenty years before LEDs came to be used in the exterior lights of a car. And even then, being red, they would be at the back of the vehicle, not the front, in the form of high-mount center brake lights. The first production car equipped with these LEDs—made by Stanley—was the 1994 Nissan Fairlady.

High-brightness LED brake lights were not only smaller, lighter, longer-lasting, and more robust than the incandescents they replaced; they were also safer. Out on the highway at sixty miles an hour, they turned on instantaneously, giving the driver of the vehicle behind a pre-cious extra car-length's worth of reaction time. At the time of this writing, high-brightness white LEDs are being used as daylight running lights in upscale European cars such as the Audi A8. When I visited Craford in mid-2005 at LumiLEDs' headquarters on the edge of Silicon Valley near San Jose Airport, he was confident that this was just the beginning. "Headlights will follow," he predicted confidently. "It's a question of, Is it two years, or three years? But certainly within this decade, you're going to have LED headlights; the whole car is going to be LEDs."

The reason for the improvement in LED brightness was the move to a more complex device, the double heterostructure. This, as we have seen, consists of two layers of slightly different material confining elec-trons and holes in a very thin layer, which in its ultimate form is known as a quantum well (a term that Holonyak claims to have coined).

The idea for the heterostructure came from Herbert Kroemer, an exact contemporary of Holonyak, who left his home in East Germany in 1948, at the time of the Berlin Blockade. In 1963, in the immediate after-math of the great race to build a semiconductor laser, Kroemer was working at Varian Associates in Palo Alto, California. There, he attended a talk on the early lasers given by a colleague. Recall that these worked only when dunked in liquid nitrogen and zapped with short pulses of current. To be useful, lasers would have to work continuously, at room temperature. This, the speaker reported, had been investigated by experts and ruled out as fundamentally impossible. "That's a pile of crap," Kroemer growled, with characteristic bluntness.

It was immediately obvious to him that all you had to do was trap the electrons by having a material with a wider bandgap confine a material with a narrower bandgap. This flash of insight would contribute to Kroemer winning the Nobel Prize for Physics thirty-seven years later. He wrote a paper on heterostructures and submitted it to *Applied Physics Letters*. The paper was rejected. (It was subsequently published, in another, less-read journal.) Then came the final irony. Varian refused Kroemer permission to work on his proposed new laser. What was the point? There would never be any applications for such a device. "There go compact discs and fiber optics," Kroemer snorted, decades later. And bright blue LEDs he might have added, since they, too, are based on double heterostructures. Today, in a nice coincidence, Kroemer works at the University of California at Santa Barbara's Engineering Sciences Building, just a few doors down the corridor from Shuji Nakamura.

It was left to others to pursue semiconductor laser technology: notably, Zhores Alferov, co-winner of the Nobel Prize with Kroemer, and Izuo Hayashi and Morton Panish, who built the first room-temperature, continuous wave laser at Bell Labs in 1970. Seven years later, Holonyak and his students used liquid phase epitaxy to grow the first quantum wells. Shortly afterward, Russell Dupuis, another of Holonyak's graduates, grew the first quantum well lasers using a new and improved growth method called metal organic chemical vapor deposition. MOCVD, as we shall see in the next chapter, was the method that Nakamura would also adopt to make his bright blue LEDs.

In November 2003, in a ceremony at the White House, President George W. Bush honored Holonyak, Craford, and Dupuis with the National Medal of Technology, the highest honor the United States can bestow on its inventors. Despite such high-profile recognition, however, the Bush administration was not prepared to put its money where its mouth was.

In 1999 a proposal to launch a large-scale US national R&D initiative on solid-state lighting had been floated. The idea was to accelerate the transition from conventional lighting to energy-efficient LEDs. In addition to companies, the initiative would involve universities and government institutes like Sandia National Laboratories. The program would run for ten years with $50 million in annual funding. Senator Jeff Bingaman, a Democrat from New Mexico, sponsored a bill in the Senate that would create such an initiative, dubbed the Next Generation Lighting Initiative. This was subsequently lumped into the Bush administration's huge, bloated, controversial, oil industry-friendly Energy Policy Act of 2005. Funding for LEDs had been slashed to a measly $5 million a year.* "That's not really enough to make an impact," Craford told me. "The Bush government is more into trying to find more oil than trying to save energy."

While it may have missed the boat on funding for research and development of core technologies, there is still plenty the government can do to accelerate the market for solid-state lighting by promoting the adoption and growth of energy-saving lights through advocacy, incentives, and rebate programs.

More on government LED initiatives in part 4. Meantime, let us return to the story of Shuji Nakamura.

---

* It has since been increased, but not by much.

CHAPTER 3

S huji Nakamura was angry. It was July 1987 and during his eight years as a salaryman at Nichia, Shuji had done everything he had been told to do. Working alone he had developed compound semiconductors, epitaxial wafers, and red and infrared light emitting diodes. But to no avail, they had all been commercial failures. At this point, his standing within the company could scarcely have been lower. He had been passed over for raises and promotions. His co-workers blamed him for wasting the company's money on his research. And, since in order to protect its know-how the company did not permit employees to publish papers, Nakamura had nothing to show for his efforts professionally, either.

It was all so unfair, he reflected bitterly. After all, it wasn't as if he had been the one who had decided to develop these uncompetitive products. The company's salesmen had him assigned to develop them. Nonetheless, he was the one who had to shoulder the blame for their failure in the marketplace. The more he brooded about it, the angrier he became. And he felt guilty toward his friends at Nichia. Simple country folk that they were, his workmates had expected great things from him, new products that would increase the company's sales. He had let them down.

Now, feeling cornered and ill at ease, Nakamura brooded about what to do next. Finally, he decided that the only way out was to quit Nichia. Every morning he would go to work thinking that, that day, he would hand in his resignation. Two things made him hesitate. One was the very practical consideration that he had a family with three young daughters to support. The other was his desire to pay back what he felt he owed the company, by developing a successful product. Only then would he be able to look his colleagues in the eye. Of course it was very unlikely that he would succeed in such an endeavor. But by this stage, Nakamura was a desperate man.

Thus far, he had not done any research that he himself really wanted to do. He had merely done what he had been told. "Since I'm going to quit anyway," he said to himself, "from now on, I should take risks and do exactly what I want to. If I fail, then that will be a good excuse to quit." His goal therefore should be as ambitious as possible, something that was almost impossible to achieve. A bright blue LED would be ideal.

He had been mulling the idea in his mind for a long time. Through his reading of the technical literature, Nakamura was well aware that no one had managed to develop a bright blue LED. Its absence was an affront to any self-respecting LED researcher, and Shuji was no exception in this regard. He had mentioned his ambition to his immediate boss several times, only to be flatly rejected. "Are you stupid?" his boss asked. "There's no way you could develop something like that. The big companies have all tried and they haven't been able to—what makes you think you could do it here, at a small company, with no budget and without proper equipment?"

Now, however, brooding about the matter, Nakamura had come to the conclusion that he should no longer listen to what his boss said. Quite the reverse, in fact: he would do the opposite of what people told him. What the company opposed was something worth doing. He also realized that there was no point telling his immediate boss that he wanted to aim at developing a bright blue LED. The latter would only laugh at his pretension. More to the point, his boss did not have the power to make decisions on such an important issue. Big decisions at Nichia were made by top management. Mostly that meant the president, Nobuo Ogawa.

Though at seventy-six he was no longer involved in the day-to-day running of the company, when it came to crucial decisions, old man Ogawa still called the shots. Nakamura knew that if he was to have any chance of chasing his dream, he would have to gain the president's approval. It was not just a question of money. Once he had secured the president's permission, no one would dare to stand in his way. Shuji was aware from what people had told him that, because of his ability to make things, Ogawa was favorably disposed toward him. Nonetheless, Nakamura was taking a big risk: he determined that if the old man turned him down, then he would have to quit. By that stage, however, he felt that he had nothing to lose by being audacious. If they wanted to fire him, let them; he didn't care any longer. Nakamura had reached the end of his tether. He was not afraid of anything anymore.

So, summoning up all his courage, at ten o'clock one morning Shuji marched into the president's office to make his case. He didn't go into detail—the old man would probably not understand it if he did—he just came out and said: "I want to develop a bright blue LED." To his amazement, Ogawa simply replied: "Is that what you want to do? Well, in that case—go ahead." Nakamura could hardly believe his ears. Concerned in case the president had misheard or misunderstood him, he sought confirmation: "Really? Is it OK if I try to develop a bright blue LED?" Then Ogawa repeated, loud and clear this time: "If that's what you want to do, then go ahead."

*   *   *

Nakamura had joined Nichia in April 1979, along with six other new employees. They were all locals, mostly from Anan; he was the only one not from Tokushima Prefecture. Shuji was assigned to the development section. To his surprise, this comprised just three staff: the section chief, one other researcher, and himself. Soon it would consist of just two, Nakamura and the section chief. Others alerted him to the fact that, at this company, the development section was not exactly the fastest of career tracks.

Nichia at the time had fewer than two hundred employees. The company's main business was the production of phosphors—materials that glow when hit by electrons. Phosphors are used in color televisions, fluorescent lamps, and x-ray intensifiers. All three were mature markets. If Nichia was to grow, then it needed to find new products. Coming up with such products was the mission of the development section, the only part of the company not involved with phosphor production.

Nakamura's first job was to refine high-purity gallium metal. This was a promising start. Gallium is a soft, silvery-white metal that occurs in trace amounts in aluminum ores. When compounded or alloyed with other elements, gallium is the sine qua non of almost all LEDs. Nakamura had worried that, at a company whose business was producing chemical powders, he would not be able to exploit his skill in electronics. Now, because he was to work on gallium, a semiconductor material, he felt somewhat reassured. Nichia had embarked on the refining of gallium metal some years before Nakamura joined. But sales were disappointing. So much so in fact that the company's managing director, Nobuo Ogawa's son-in-law Eiji, was considering disbanding the development section altogether.

After Nakamura had been at Nichia a couple of months, the company issued instructions that the development section was to move on to a new project. Nichia's salesmen had picked up a promising hint during a visit to a big customer in Osaka. Refining gallium might be a dead end, but if they could go one step further and produce gallium phosphide, a compound semiconductor material used to make red and yellow-green LEDs, they could sell ingots of that instead.

Nakamura was relieved that the development section was to be spared. He would have been less so if he had understood what was really going on. Namely, that the customer was just looking to outsource the production of gallium phosphide to a cheaper supplier. Nakamura had worked on compound semiconductors during his master's course. It was natural that he should be assigned to the project. He began research, throwing himself into the work with great determination. Knowing next to nothing about growing LED materials—which comes under the heading of chemical, rather than electrical engineering—he had to start from scratch. Since he did not get on with the section chief, he had to work mostly on his own.

Shuji spent six months reading the literature of LEDs, immersing himself in the theory, a task that he greatly enjoyed. But when the time came to do experiments, Nakamura discovered that, just like at Tokushima University, there was no budget for equipment. He had to make do by scavenging obsolete bits and pieces that were lying around the place, fixing broken parts, cobbling together everything he needed by hand.

To make gallium phosphide, you place your starting materials—gallium metal (which, like mercury, is liquid at room temperature) and phosphorus—at either end of a sealed, evacuated quartz tube. You heat them up above their melting point, causing them to react with each other. After the reaction is complete, you gradually reduce the temperature. A D-shaped ingot of single-crystal gallium phosphide forms. Finally, you cut off the end of the tube with a diamond saw and remove the ingot.

To build his equipment, Nakamura scrounged some heatproof bricks, cables, a vacuum pump, and an old electric furnace that had formerly been used to make phosphors. The quartz tubes he had to order. Quartz is necessary because it can withstand high temperatures (glass would just melt). But the tubes came open-ended. In order to seal them so that they could be evacuated, he had to learn how to weld quartz.

The production process involves heating up the tube until it reaches about 1,100 degrees Celsius at the gallium end and about 600 degrees Celsius at the phosphorus end. At such temperatures, the tube glows bright red. Controlling the temperature is crucial, because if it gets too high, the phosphorus vapor expands, causing the quartz to crack. This lets in oxygen, which reacts with the phosphorus, causing an explosion. Such explosions became a feature of Shuji's life at Nichia. Following them, the whole room where he was working would instantaneously fill with dense white smoke. The phosphorus, which had ignited, would fly everywhere, all over the walls and the floor, along with shards of broken quartz tube. Nakamura would run around desperately pouring water over the burning phosphorus, trying to douse the flames, which sometimes reached as high as the ceiling.

The blasts happened several times a month, typically around five o'clock in the evening. The shock wave from the explosion would hit his fellow workers as they were heading for their cars in the parking lot about a hundred yards away. The first few times it happened, they dashed into his lab shouting, "Nakamura—are you still alive?" By the fifth or sixth time, however, they had become so used to the bangs that they no longer came to check on his well-being. They would just say to each other, "Sounds like he's done it again."

After the initial explosions, Shuji constructed a primitive protective aluminum shield about three feet square. He erected it in front of the electric furnace in order to contain the blast. Years later, Nakamura would shake his head in disbelief as he recalled how reckless he was and how dangerous his circumstances had been. And he would ask himself why Nichia had been so negligent about the safety of its employees.

The explosions were a setback in another way, too. Whenever equipment broke, Nakamura would have to rebuild it. He became expert at welding quartz tubes so that they could be reused after the ingots had been cut out. Some days he would spend from eight in the morning until three in the afternoon just welding tubes. It was hot and sweaty work, using an oxyacetylene burner. It was also tense, because any crack in the weld would cause an explosion. As a graduate student, Shuji had felt like a worker in a sheet metal factory. Now, it seemed, he had become a welder. He felt his life as a researcher had ended before it hardly even had begun.

All the time, he had to endure the company putting pressure on him. "Haven't you made anything yet?" his boss would say. "Commercialize a product quickly and contribute to sales." Eventually, after about three years' effort, Nakamura succeeded in developing commercial-grade gallium phosphide. More satisfying than producing the material, however, was how he felt when the company's salesmen reported back telling him that they had made a sale. It gave him pleasure to think that he had finally managed to contribute something to the company's bottom line. But only a little: the market for gallium phosphide was already crowded with suppliers. As a late entrant, Nichia was only able to win a sliver of the pie. Monthly sales of gallium phosphide, a relatively low-cost material, never amounted to much more than a few tens of thousands of dollars.

※   ※   ※

Nakamura's next assignment was to produce gallium arsenide. Like gallium phosphide, gallium arsenide is used to make LEDs, typically infrared ones such as those found in television remote controls. But GaAs also has other applications, such as the semiconductor lasers used in fiber-optic communications. Thus, the potential market for the material was larger.

The manufacturing method was the same as for gallium phosphide. Once again, during the experimental stage, there were many explosions. Happily, unlike phosphorus, arsenic is not inflammable, so it was easier to work with. Unhappily, the material is highly poisonous, releasing lethal arsenic oxide gas every time the furnace blew up. Nakamura would wait until the smoke cleared before going to clear up the mess. He had to wear a homemade "space suit" and breathe through a respirator. By some miracle he was never adversely affected by having to work in such a toxic environment. And all the time he was learning the "black art" of making compound semiconductor materials, the kind of detailed know-how that scientific papers never mention. For example, how to control the temperature of an electric furnace with exquisite precision. His masterless apprenticeship would be invaluable for his later work in developing the bright blue LED.

By 1985 Shuji was producing gallium arsenide in bulk quantities. Looking back, he was inclined to think that this was quite an achievement. A big company would have assigned a team of people to such a project. It is not unusual for development to take five years or more. And yet he had done it in just three years, all by himself. But when it came to selling the product, the market's response was the same. There were plenty of existing suppliers, so why would customers buy from an untried Johnny-come-lately like Nichia?

The next bright idea the salesmen brought back was, instead of making the starting materials for LEDs, why not go the whole hog and make the wafers and light emitting devices themselves? To fabricate a simple LED required mastering a technique known as liquid phase epitaxy, in which successive layers of crystalline material, each with different electrical properties but identical crystal lattices, are deposited on the wafer, one after another. Chop the wafer up into chips, stick electrodes on either end of the chip, and bingo! You've got an LED.

To learn how to do liquid phase epitaxy, Nakamura had to resort once again to the technical literature. Having pored over journal articles and published patents, he performed countless experiments, varying the temperature at which the layers were grown and the time it took to grow them. Small differences in thickness, he discovered, could make a big difference in brightness and lifetime. As usual, there was pressure from the company to produce a saleable product quickly. As usual, there was no budget available to buy equipment. Eventually, by repeated trial and error and making improvements based on the results, he managed to make some prototype LEDs.

Samples of this device were delivered to a client for evaluation. The client responded that the device wasn't very bright and degraded over time. Not having measuring equipment of his own meant that Nakamura was dependent on such evaluations from clients. He had to wait several months to get data back before he could start making improvements. Shuji felt strongly that if the company was going to enter the LED device business, then he should be able to conduct his own evaluations. He tried to argue the case with his boss, but was told, No budget, so not possible.

Previously Shuji would have accepted this answer and given up. By now, however, he had been with the company for more than six years. He was in a position where he could make his voice heard. Also, he had come to realize that Nichia was a company that was run on the say-so of its president and founder, Nobuo Ogawa. He decided that it would be more effective to approach the top man. Shuji went to Ogawa and asked him for what he needed. To his surprise, the old man agreed immediately, giving his permission for Nakamura to buy the equipment to measure luminous intensity and device lifetime. This encounter set a precedent, one upon which Nakamura would draw three years later, when he went to see Ogawa again, this time with a much bolder request.

By 1987 Nakamura had developed commercial-grade infrared and red LEDs. He had done everything from R&D, to manufacturing, to quality control. Now, the company tasked him to add yet another string to his bow: sales. Since Nichia did not have any salesmen who under-stood semiconductors, it fell to Nakamura to go out and visit customers, large consumer electronics manufacturers that used LEDs in their prod-ucts, and to try and sell them the devices he had made. Often he would visit the factories of Matsushita and Toshiba, Nichia's biggest cus-tomers. There, he would meet researchers with PhDs whose speciality was semiconductor technology. The experts would raise their eyebrows at the samples he brought. "How come someone like you is able to make semiconductors all by yourself in a hick town like Anan?" they asked him. "Especially without introducing technology from somewhere else?" Buoyed by such compliments, Nakamura began to acquire confi-dence in his abilities as an LED researcher.

Potential customers would also come down to visit Tokushima. Nakamura and Nichia's salesmen would wine and dine them in local clubs and karaoke bars. Shuji spent so many nights entertaining that he began to worry he might be permanently reassigned to the company's sales department. Sociable and friendly by nature, Shuji was popular among the salespeople. He would probably have made an excellent salesman. But no matter how persuasive his pitch, customers were still reluctant to buy LEDs from a company that had little or no track record in the semiconductor business. There were plenty of other, much better-established suppliers in the market. Inevitably, customers asked for reductions in price, one company demanding a discount of 50 percent. It is sometimes said that the Japanese economic miracle was built on the sacrifice of small subcontractors.

Nakamura could never manage to raise sales of LEDs beyond a few tens of thousands of dollars a month. Not much, considering the expen-sive equipment for measuring, testing, and production that he had per-suaded the company to buy him. Or the salaries for the new employees that joined the company each time he had commercialized a new product to do the manufacturing. Even though, as Nakamura realized, because he had done all the product development himself and custom-built much of the equipment, the actual development costs were, by industry standards, minuscule.

Nakamura had made many friends among Nichia's employees. When work finished at five, they would often ask him to make up the numbers for a game of baseball or softball. Afterward, they would all drop by a local bar. After a few drinks had loosened their tongues, his workmates would say to him, "Nakamura-san, develop some good products and make the company grow. We're just country people, farming is all we know, but we have faith in you." And feeling small, knowing that he had yet to make anything that had an impact on the company's bottom line, Shuji would hang his head. Others, especially older employees, were critical. On an overnight business trip to Tokyo, after a few drinks, Nichia's sales manager started haranguing Shuji. "The development section is just a name: What have you been doing for the past five years? We can't sell what you've developed—you're just wasting the company's money." Feeling miserable and unable to come up with a rebuttal to these charges, Nakamura just kept apologizing until the sales manager fell into a drunken sleep.

His situation was truly depressing. The fact that he had made sophisticated semiconductor products essentially unaided should have been appreciated. In fact, he had been repeatedly passed over for promotion. In part, this was because he had never been involved in manufacturing, the section that typically garners most praise, because that is where sales derive from.

Whenever it came time to commercialize a product, the company would employ someone to take charge of the manufacturing. Shuji would teach him what to do, then he would be assigned his next project. This latecomer, who had not contributed anything to product development, would be promoted ahead of him to become his boss. He felt as if the credit had been stolen from him. And he remained on the lowest rung of the salary scale. The only way a corporate researcher can contribute directly to the bottom line is through patent royalties. But fearful of losing precious trade secrets, Nichia did not permit patent applications on principle. Thus Nakamura's apparent sales were zero.

Elsewhere, too, he felt invisible. Few people in the industry even knew that Nichia made semiconductor materials and devices. Sometimes he would call semiconductor equipment makers and ask them to send catalogs. Perhaps feeling that an inquiry from a remote rural location like

Anan was unlikely to lead to a sale, the suppliers would not bother to send them.

In ten years Nakamura had not published a single scientific paper, because of Nichia's policy of keeping its technical know-how a tightly guarded secret. Nor had he ever been to a conference. From a professional point of view, he had no achievements, he did not even exist. He had been a loyal and obedient employee, and where had it got him? Now, in early 1988, having won the president's consent to go ahead with his audacious plan to develop a bright blue LED, Shuji decided it was about time he made his presence felt.

*   *   *

Not for the last time in this story, Shuji's timing was impeccable. During the 1980s, riding the boom in personal computers, which meant increased demand for monitors, hence phosphors, Nichia's sales had been rising steadily. The company was making good money. Nonetheless, president Ogawa was taken aback when Nakamura returned to see him to explain that he would need around three hundred million yen (worth in 1988 about $2.4 million). This was equivalent to 2 percent of the company's sales that year. Compared to previous research costs, it was an unbelievably large amount. "That much? I see," Ogawa said. "Well, let me think about it." Nakamura went back several times, explaining about the bright blue LED, why it would be a big market, and why it would cost so much to develop. Eventually, the old man gave Nakamura his blessing.

In a 1997 book he coauthored with Nakamura, Gerhard Fasol put the figures into perspective. "It is rare for a large company to spend [$2.4 million] within essentially one year on a single blue-sky type research project of a single researcher," Fasol pointed out. "It is even rarer that [2 percent] of annual sales would be spent on a single blue-sky project of initially unknown outcome, such as Nichia [did]." Large R&D-based companies like IBM spend an average of around $250,000 to $400,000 per researcher. Their approach is usually to provide small budgets until commercial success is within reach, or at least until the risk is easier to assess.

In fact, Nakamura's apparently quixotic quest to develop a bright blue LED would end up costing Nichia even more than originally budgeted. By 1990 the company had spent five hundred million yen (about $4 million) funding Shuji's work.

Nakamura had calculated how much money he would require based on the expertise he needed to acquire and the equipment he needed to buy. Two-thirds of the money would go to equipment, together with the laboratory and the clean room facilities to house it. Of the remaining third, one large expense item was time to be spent in the United States, mastering the crystal growth technology that Nakamura reckoned he would need in order to make bright blue LEDs. This was called metal organic chemical vapor deposition, MOCVD for short.

The conventional method of making LEDs, liquid phase epitaxy, could not grow films of high enough quality and sufficient thinness. There were two more modern alternatives. Both were capable of growing films of material just a few atoms thick. One, molecular beam epitaxy, works via vacuum evaporation. MBE is especially popular with academic scientists. It can produce small quantities of material, enough for researchers to extract data based on which they can write and publish papers. But MBE requires an ultrahigh vacuum, has very slow growth rates, and is difficult to scale up. In the opinion of most people, Shuji included, the method is not suitable for mass production.

That left MOCVD, which does not need a high vacuum and can be applied to the factory floor. The choice was thus, as they say in Silicon Valley, a no-brainer. Shuji selected MOCVD without hesitation. But he had little idea of how MOCVD was done. By a stroke of good fortune, it just so happened that one of Japan's leading experts on the technique was an old acquaintance of his from Tokushima University. Though Shiro Sakai had been two years Shuji's senior, they had worked together in the same laboratory, and Shuji knew him well. In the intervening years, Sakai, now a professor at Tokushima, had become well known for his expertise in MOCVD. Now he was on sabbatical at the University of Florida. During the summer holiday of 1987, he returned to Japan. Nakamura went to see Sakai to ask his advice on how to learn MOCVD.

When Sakai returned to Japan for a week at the end of 1987, Nakamura invited him to visit Nichia. There, the professor explained to Ogawa the significance of MOCVD as a crucial tool for the production of state-of-the-art LEDs. At this meeting, blue LEDs were not mentioned. Sakai recommended that Nichia should send Nakamura for a year to the engineering school at the University of Florida, where he was currently on sabbatical. A deal was arranged: Nakamura would learn MOCVD under Sakai's tutelage. In return, Nichia would donate around $100,000 to fund Sakai's research.

*   *   *

To make a chip the size of a grain of sand takes a mighty big box. A typical MOCVD system is almost as big as a shipping container and costs well over a million dollars. Seen from the outside, MOCVD equipment looks rather dull, like a row of office cabinets. Peek behind the bland-looking doors, however, and you will discover a bewildering assemblage of tanks, pumps, and valves connected by what appears to be several miles of thin, stainless steel pipe. At one end of the cabinets is a rack containing a computer that runs the recipes for growing LEDs. These are programs that, with exquisite precision, control the pressure and flow of gases, while monitoring the temperature and the rate at which the thin films of crystal grow.

The heart of an MOCVD system, visible through a little window in one of the cabinet doors, is its reactor chamber. This is a cylinder about the size of a cookie jar, made of quartz in some systems, of metal in others. It may be positioned either horizontally or perpendicularly. The chamber is remarkably small in comparison to the whole. It occupies perhaps 3 or 4 percent of the total space. Inside the jar is a graphite chuck mounted on a little pedestal. Here sits the wafer on which the thin films are grown. The chuck is connected, via a thermocouple that monitors the growth temperature, to a heater. To grow gallium nitride, the wafer is heated to between 1,000 and 1,200 degrees Celsius, at which point it glows bright golden-orange. In the case of quartz chambers, the heat comes from copper coils wound round the jar. An exhaust system, typically a vacuum pump, completes the process. It sucks the unused gases out of the reactor chamber, flushing them away to a scrubber for disposal.

For more than twenty years LEDs were grown by one of two methods, liquid phase epitaxy (LPE) or vapor phase epitaxy (VPE). Epitaxy simply means stacking crystal layer upon crystal layer with exactly the same orientation, like piles of egg trays. But when it came to growing high-quality thin films and quantum wells, which require abrupt atomic-level transitions from one layer to the next, both processes were too crude. For example, an LPE system consists of a quartz tube in which are lined up little graphite dishes called, because of their cigarlike shape, "boats." Each boat contains a different semiconductor material that is heated until it melts. You slide your wafer along the tube, leaving it to sit a while on top of each boat. Cooling causes some of the material to precipitate onto the surface of the wafer. LPE produces relatively thick layers, and the boundaries between them are gradual rather than sharply defined. Precise control over thickness is almost impossible to achieve.

MOCVD (sometimes, confusingly, also known as MOVPE) systems became the method of choice for growing high-brightness devices, originally red LEDs, in the mid-1980s. MOCVD accomplishes the abrupt transition between layers by allowing the crystal grower to run two mixes of gases through the system simultaneously. While using mix A to grow a film in the reactor, you have all the gases for mix B flowing directly to the exhaust. Then, at just the right moment, you switch mix A to the exhaust and send mix B to the reactor. The process takes a matter of moments. All you hear is the sound of the compressed air-driven pneumatic valves. They open and close in quick succession—phsst, phsst, phsst, phsst—*et voila!* You have grown a quantum well.

So much for vapor deposition. Now we come to metal organic chemicals. Why it is necessary to use such fancy-sounding stuff instead of ordinary metal? The answer is that, in their vapor phase, neither aluminum, gallium, nor indium—the three most common metals used in growing bright blue (and red and green) LEDs—can muster sufficient vapor pressure to make it as far as the jar under their own steam. They have to be picked up and carried there, in organic form. To bump up the vapor pressure, organic chemicals such as methyl groups are attached to the metals.

Gallium becomes trimethyl gallium; the positive-type dopant magnesium becomes bis(cyclopentadienyl) magnesium, mercifully abbreviated as CP2Mg. The carrier gas is hydrogen. It is kept flowing through the system at the rate of many liters per minute. During its travels, the hydrogen bubbles through the temperature-controlled baths that contain the metal organic mixtures. The gas picks up some of the compounds, which it transports to the reactor. When the compound gases get to the hot zone, they lose their methyl groups. The nitrogen for gallium nitride arrives at the jar in the form of ammonia. The heat decomposes the gas, leaving nitrogen atoms hot to trot with their gallium partners.

The process of growing a gallium nitride LED begins by heating the sapphire wafer to a very high temperature. Once hot, you clean the surface by flowing nitrogen over it. Then you drop the temperature way down to maybe 500 degrees Celsius to grow the first layer, the so-called nucleation, or buffer, layer. This is a thin film, typically of gallium or aluminum nitride and just fifty to one hundred atoms thick, that is grown directly on the wafer. The buffer layer is amorphous, that is, lacking a crystalline structure. When you heat it up, the surface of this amorphous layer becomes very lumpy as nucleation islands oriented to the surface of the sapphire start to form. As you reach higher temperatures, however, these islands grow together laterally, to form a smooth, mirrorlike layer of gallium or aluminum nitride. One of the secrets of growing high-quality GaN is being able to control exactly how this nucleation layer is deposited, how it crystallizes, and how it grows together during the heat-up step.

On top of the nucleation layer, you deposit plain vanilla (i.e., undoped) gallium nitride. Next comes a layer of negative-type gallium nitride, with silane as the electron-donating dopant. That is followed by a layer of negatively doped aluminum gallium nitride, a compound with a wider bandgap than GaN. This layer plus another, positively doped layer of AlGaN on the other side serve to confine the charge carriers within the active (i.e., light emitting) layer of the device. Then you drop the temperature down from 1,000–1,200 degrees Celsius to 750–850 degrees Celsius so that you can grow an indium gallium nitride quantum well. You grow, say, 20 angstroms of InGaN, then maybe 100 angstroms of GaN, then repeat the process for as many quantum wells as your recipe calls for, adjusting the amount of indium to produce the desired wavelength of light. The more indium you include, the greener the output will be. After growing your last combo of InGaN + GaN, you crank the temperature back up and deposit your other confining layer of positively doped aluminum gallium nitride. Then you cap the whole thing off with a layer of positive-type gallium nitride using magnesium as the hole-donating dopant. That completes the device.

In a typical growth run, the whole process takes anywhere between two and a half and four hours. If you load your wafer first thing in the morning just as the coffee is brewing, you will get the growth run out around lunchtime. You could schedule another run around two o'clock and have it out before dinner. Between runs, you have to clean the reactor by baking it out at high temperature. In the R&D lab, two runs is not a bad day. On the production line, four growth runs in a twenty-four-hour period is considered pretty good going. A large production-line reactor may contain a platter with as many as one hundred wafers on board.

The growth process itself is not in the least dramatic. You can hear faint hums and hisses from the pumps and the valves, but that's about it. The only smell MOCVD machines give off is a subtle whiff of burnt reactants that emanates from inside the jar. If you smell anything else—ammonia, for example—that means there is a leak. This is a good time to leave the lab. Quickly.

Though in many ways the epitome of high tech, operating an MOCVD reactor is actually a fine art. Herb Maruska, a gallium nitride pioneer we shall meet in the next chapter, compares it to playing a musical instrument. "Lots of people can play the violin, but only a few have the superior musical abilities that make them virtuosos. Similarly, some people really understand the peculiarities of their reactors: they know just how to position the wafer, to set the flows of various gases, to switch the temperatures. It really comes down to feel."

＊　　＊　　＊

In March 1988, three months after getting the green light to go ahead with his plan, Shuji was on a Delta Airways jet flying from Tokyo via Atlanta to Gainesville, home of the University of Florida, the fourth-largest public university in the United States. It was the first time the country boy had ever boarded an airplane. Like many first-time fliers, Shuji fretted that the plane might fall from the sky. It was also the first time he had been abroad. Assuming he did manage to arrive alive, he was worried that his rudimentary and thus-far-untested high school English would enable him to communicate with Americans.

As it happened, language would not be Nakamura's main problem in the United States. Most of his fellow students in Ramu Ramaswamy's laboratory were also from Asia, mainly Korea and Taiwan. English was not their first language, either. Shuji was astonished by the ethnic diversity of the student body at Gainesville. Which country does this university belong to? he wondered. In later life, Nakamura would marvel at the generosity of Americans in inviting knowledge-seekers from all over the world to come study on their campuses. Such openness compared favorably, he felt, with the close-mindedness of their Japanese counterparts.

In 1988 Shuji turned thirty-four, rather long in the tooth for a student. His fellow researchers were mostly in their mid-twenties. All of them had doctoral degrees. A common failing among academics, especially young academics, is that they tend to be overly status conscious. A person with a PhD will typically let you know that he has a PhD within moments of your first meeting. This failing is particularly pronounced among Asian academics, whose hierarchical cultures produce an exceptional degree of status consciousness.

Shuji's status at Gainesville was ambiguous. He was coming to the university at the behest of Sakai, a visiting professor. Since his time was limited to one year and he was not studying for a degree, he was obviously not a student. Nor, since he did not have a PhD, could he be offered a postdoctoral fellowship. As a compromise, he was designated a "guest research associate."

Nakamura arrived in Florida as something of a mystery man. Nothing was known about him other than what he had written about his previous work experience and his proposed research theme. This of course did not include any mention of bright blue LEDs. Formally, he was there to do research on infrared LEDs made of gallium arsenide.

Nichia's real intentions in sending him to the United States would remain a closely guarded secret.

Initially, his colleagues treated Shuji as an equal or even, because he was older, as a senior. However, once they discovered that he had only a master's degree to his name and, worse, that he had not published a single paper, their attitude toward him changed completely. Henceforth they looked down on Nakamura, treating him as little more than a lab technician. Nakamura felt humiliated. It was particularly galling because, from his perspective, these puffed-up PhDs were mere novices. He had years of hands-on experience under his belt. They could not do the simplest experiment without making a fuss. Something would go wrong and they would come crying to him for help. He would show them what to do. But that did not make the brats any less snooty in their attitude toward him.

For Shuji, Fridays were sheer torture. On that day, Ramaswamy held discussion sessions that lasted from eight o'clock in the morning sometimes until quite late in the evening. "We'd talk about the research papers they were writing," Ramaswamy recalled, "with every student going to the board and discussing their concepts, their problems, how things could be modified, and this and that. I got the feeling that Nakamura was intimidated, because he hadn't published anything, didn't have a PhD. I used to look at his face and he would not be very happy, he would look very perturbed. He didn't have the confidence, he wouldn't ask any questions, he was very shy and [perhaps because of the language issue] he spoke very little."

Nakamura was by nature a diligent worker. The guilt he felt about his failure to develop a commercially successful product combined with his anger at the way Nichia had treated him served to motivate him further. He had always hated to lose. Now, the arrogance of these greenhorn academics poured fuel on the competitive fires that burned within him.

"I do not like to be defeated," he wrote. "I feel resentful when people look down on me. At that time, I developed more fighting spirit—I would not allow myself to be beaten by such low-level people."

In a word Nakamura was, as Ramaswamy put it, *driven*.

"He was a bulldog worker, he would work around the clock. I used to come back to the lab late at night, sometimes I'd stay until ten, then I'd go home and I'd forget something. I'd get in my car and drive back to the lab and I'd see him working at two o'clock. Next day I'd come back at five or six in the morning, and he'd be still there! I'd say to him, Don't you go home and sleep? And he'd say, Well, I was in the middle of this, so I thought I'd finish it. I think maybe he felt somewhat insecure, because he was so motivated, and so driven."

Outside of his work at the laboratory, not much distracted Shuji during his time in Florida. Gainesville was not much to his liking. "The end of the South," as the town is sometimes known, the birthplace of the sports drink Gatorade was, as Nakamura saw it, "surrounded by a swamp full of alligators and mosquitoes, a place where African Americans were still discriminated against."

He lived in three-hundred-dollar-a-week student accommodation, eating out most nights at a cheap Chinese restaurant nearby. His only friend was a Japanese professor who lived in the neighborhood. The kindly academic would invite his fellow countryman to go fishing with him, then back to his apartment for dinner where he would cook what they had caught. On this first long absence, Shuji was naturally homesick for his family and his native land. During the summer holidays Hiroko and the girls came to visit him. He treated them to a trip to Disney World in nearby Orlando. During his year in the United States, it was the only time he took off.

In addition to wrestling with his inner demons, there was also a pressing practical reason why Shuji worked so relentlessly. He had come to Gainesville to learn how to do MOCVD. When he arrived, however, he had been shocked to discover that getting time to experiment on the university's two existing MOCVD systems was not going to be easy.

State-of-the-art equipment is always fully booked. Inevitably, the lion's share of access time goes to the most powerful. "Professors are like . . . how shall I put it? Wild animals roaming the mountains," Ramaswamy said. "You can't get them in a corral and make them go round and round." Others saw the situation more prosaically. A turf war had been waged over control of the systems at the university, and it seemed like the gentle Ramaswamy had lost out.

Not surprisingly, Ramaswamy wanted a machine of his own for his lab. Commercial MOCVD equipment did not quite suit his purposes, so he had brought in Sakai to custom design a system for him. By the time Nakamura got there, the parts had arrived, but they had yet to be assembled. Shuji realized that if he was ever going to have a chance of using this machine, then he would have to help Sakai build it. He had to spend ten months of his precious year in the United States with his sleeves rolled up, connecting pipes and welding quartz, just like back at Nichia. He threw himself into the task, working sixteen hours a day, seven days a week.

Here again, as with Shuji's long and apparently fruitless apprenticeship learning the basics of LED growth, adversity in the short term would turn out in the long term to be priceless training for his quest to develop the world's first bright blue LEDs. He was willy-nilly gaining an intimate familiarity with the inner workings of MOCVD equipment that few could match.

At last, having managed to assemble the equipment, Nakamura was not about to waste what little time in the United States he had left. He wanted exclusive use of the machine that he had done so much to assemble. This led to clashes with the other students. Ramaswamy was forced to intervene to settle the issue. In the end, Nakamura was only able to do about ten device-growing runs on the system. Driven as never before, he was frantically busy right up until the last moment. Then it was time to go home.

*   *   *

Nakamura returned to Anan in March 1989. While in the United States he had ordered his own basic MOCVD equipment, keeping his goal a secret from the supplier, Japan Oxygen. He stated on the order form that his purpose was to grow gallium arsenide infrared LEDs. Now the huge $1.6 million machine had arrived, but the big question was, what material would he choose to grow in it?

There were, as Shuji knew well from his extensive reading of LED literature, and from conferences he had attended during his time in the United States, three candidates. All were compound semiconductors. Silicon carbide, despite the fact that it was already in limited commercial production, was one he had already rejected. LEDs grown from silicon carbide produced a weedy blue light, rather like the color of denim jeans that have been repeatedly washed. Silicon carbide had an indirect bandgap. In plain English that meant the material would never be able to emit bright blue light.

The other two materials, zinc selenide and gallium nitride, both suffered from the same major deficiencies. One was that, in order to make a proper LED, you need to be able to fabricate two types of material: negative type, doped with impurities to give it an excess of electrons, and positive type, doped with impurities to give it an excess of holes. Thus far, however, it had proved impossible to fabricate either p-type zinc selenide or p-type gallium nitride. At the time he was making his decision, Nakamura could not have known it, but this was about to change: in 1989, researchers would succeed in fabricating p-type gallium nitride; the following year would see the first p-type zinc selenide.

The other, more serious, drawback was the lack of a suitable base material on which to build your LED. Gallium arsenide LEDs could be grown on gallium arsenide wafers sliced from ingots of GaAs. But nobody had been able to grow bulk zinc selenide or gallium nitride. That meant you had to employ wafers of some other material as the substrate. Which in turn meant that there was always going to be a mismatch between substrate and light emitting layers. That is to say, there would be defects such as cracks in the crystal lattice. It was a bit like trying to fit Lego bricks onto a base made by another toy brick company. Defects are bad news for LEDs because they cause devices to dissipate energy in the form of heat instead of light.

With zinc selenide, a very soft material, the problem seemed much less severe. You could grow zinc selenide on a gallium arsenide substrate, and the mismatch was only 0.3 percent, not so far off the ideal value of 0.01 percent. This translated into a defect density of around one thousand per square centimeter.

With gallium nitride, a rock-hard material, it was an altogether different story. The best available substrate for GaN was sapphire. But even sapphire produced a huge mismatch of 16 percent. That translated into a defect density of a whopping ten billion per square centimeter. It was plausible to imagine that, given time and effort, imperfections in crystal ZnSe could be reduced by an order of magnitude. But *ten billion* defects? There was no way that figure was going to be significantly reduced in any researcher's working lifetime.

Gallium nitride, as we shall see in the next chapter, had been thoroughly investigated by RCA, Bell Labs, and Matsushita. GaN, or "gan" as it is often called these days, was almost universally perceived to be a dead end. Indeed, worldwide, there were only three or four groups still active in the GaN field, most of them at universities. By the late 1980s, the overwhelming consensus in the research community was that if you wanted to do bright blue, zinc selenide was the way to go. Witness large-scale programs dedicated to developing devices made from the material at universities including Brown, North Carolina State, and Purdue and at companies such as Sony, Matsushita, Toshiba, IBM, and 3M. At the domestic Japanese compound semiconductor conferences Nakamura attended, he noted that participants at the ZnSe sessions numbered in the hundreds. At the sessions on gallium nitride, only a handful of researchers would show up.

Yet zinc selenide devices, such as they were, displayed (and would continue to display) a depressing tendency to fall apart when zapped with current. For a crystal lattice, giving birth to a photon is a stressful event. Zinc selenide simply wasn't robust enough to cope with the stress. What few people could have foreseen in 1989 was that gallium nitride would turn out, for reasons that are even today not well understood, to behave very differently than any previous light emitting material. Any other semiconductor with that density of defects would be dead in the water. Much to everybody's surprise, however, defects just didn't seem to matter with gallium nitride. It would prove a magical material.

Having arrived at what he described as this "fateful fork in the road" just before his return to Japan, Shuji decided to go with gallium nitride. His reason for making this apparently reckless choice was not because he was confident that he could do what no one else had done and make bright blue LEDs. Rather it was because he had repeatedly had the bitter experience of developing products only to find that his company could not sell them because big competitors were already well established in the marketplace. If he chose zinc selenide, since big companies had had several years' start on him in developing the material, it was likely that history would repeat itself. It was already too late. With gallium nitride, however, in the highly unlikely event that he did succeed, there would be no competition, because as far as he knew no other companies were working seriously on GaN. Nichia would end up in sole possession of the marketplace. And that, remarkably, is exactly what came to pass.

Nakamura was able to make this seemingly foolhardy decision by himself without reference to Nichia's senior management because none of them knew anything about semiconductors. All they knew was that Nakamura's target was to develop a bright blue LED. The choice of methodology to adopt and the material to work on was thus his, and his alone. If he had been working at a large company, his proposal to work on a material known to be a loser would undoubtedly have been shot down before it left the ground. But as he himself would later say, "Breakthroughs are born out of unusual circumstances."

When Shuji embarked on his harebrained quest, in addition to his attempt to build a bright blue LED, he also had a second, more personal goal in mind. Even if he was only able to scrape together some basic results, at least he would have something to show for his efforts in the form of research papers. He was determined to prove to those swellheaded PhDs back in Florida that he, too, could write papers, despite the fact that doing so would mean breaking Nichia's regulations. By this stage, however, Shuji had had enough. He no longer cared about company rules. He was resolved to write as many papers as he could. With a bit of luck, he might even get his doctoral degree.*

This time, however, there was at least a chance that he would not have to work entirely on his own. He went to see Shiro Sakai, his friend and MOCVD mentor, who had himself recently returned from Florida to his chair at Tokushima University. Shuji suggested to the professor that the pair should collaborate on gallium nitride research. But Sakai rejected the younger man's proposal out of hand. He told Nakamura that for an

---

*He did, from his alma mater, Tokushima University, in 1994.

academic such as himself, it was vital to keep publishing papers. Who knew whether anything would come of gallium nitride research, whether it would generate the kind of results that could be written up and published in journals? Better to stick closer to the mainstream, where the outcomes were more predictable.

Several ironies here. First, starting in 1991, Nakamura would publish the first in a long string of papers. Within a few years of their publication, these papers would become among the most cited pieces of scientific literature in the world. Second, perhaps getting wind of the fact that Nakamura was making great progress, Sakai would come to have second thoughts about the merits of gallium nitride research. He would secretly switch his focus at Tokushima to work on GaN, without telling his former friend and colleague that he had done so. Nakamura would only find out about what was going on in Sakai's lab later, when Nichia hired some of his students. Ultimately, Sakai would even start his own GaN-based company, Nitride Semiconductor. But he was an academic, not an entrepreneur, and the company would not be a success.

In the meantime, Sakai's refusal to collaborate meant that Nakamura, once again, would have to proceed on his own. Shuji was used to that, indeed he found it easier to solve problems if he did all the thinking, without interference from others. But on his lonely marathon, he would soon come up against his most difficult challenge yet. And this seemingly insurmountable obstacle would come, not from the technology, or external rivals, but from within his own company.

But before delving further into Nakamura's travails, let us first turn our attention to the owners of the shoulders Shuji had chosen to climb upon, and the history of the development of the bright blue LED.

CHAPTER 4

T he research that would culminate a quarter of a century later in Shuji Nakamura's bright blue light emitting diode began in earnest on May 13, 1968. The date can be pinned down with unusual precision. When Herbert Paul Maruska, a twenty-four-year-old researcher at RCA's David Sarnoff Research Center in Princeton, New Jersey, went downstairs to the center's library to make copies of the prewar German papers on gallium nitride, he used pages from the recycle basket on which that date is stamped.

Over the next five and a half years Maruska and his boss, Jim Tietjen, his mentor, Jacques Pankove, and his colleague, Ed Miller, would lay the foundations for all subsequent gallium nitride blue LED research. In due course, the American researchers would—unwillingly and unwittingly— pass the baton to a Japanese, Isamu Akasaki. For the next two decades Akasaki and subsequently his student, Hiroshi Amano, would keep plugging away. Then, with the finishing line in sight, they would be overtaken from out of nowhere by Nakamura.

Herb Maruska was born of German immigrant parents during World War II in an internment camp in Texas, where he was listed as an enemy alien. Americans of German descent were rounded up and imprisoned in the same camps as ethnic Japanese. Thus, unusually for an American born in America, Maruska started life surrounded by Japanese. He grew up at the top end of Manhattan Island, in a district that is today known as Inwood Heights. In those days, the neighborhood was racially divided. East of Broadway, it was Irish; west of Broadway, Jewish. Though Maruska's parents were lapsed Catholics, the gangs of Irish kids that roamed the neighborhood assumed that the youngster was Jewish, and accordingly beat him up on a regular basis.

Unlike your stereotypical, brash New Yorker, Maruska is a soft-spoken, gentle soul. A bright kid, he attended the elite Bronx High School of Science. He filled the family's tiny three-room apartment with the tell-tale signs of the budding engineer: old radios, televisions, oscilloscopes, and voltmeters. Though his parents were poor he managed to make his way through New York University, which at one time was a private school, by winning scholarships and working part-time. He graduated in 1965 but stayed on for another year to earn a master's degree.

Maruska's goal was to design and build radios and televisions. That meant taking courses in science and engineering. With the Vietnam War under way and JFK's race to the moon in full swing, it was a time of endless job opportunities for engineers. Recruiters came flocking to campus. "As long as you were alive, you were going to get a job offer," he laughed. Maruska went on many interviews at large aerospace companies. At most places they would take the interviewees on tours. They wound up in a large room, with engineers sitting at desks, facing the front where there was a group head staring back at them. It looked a lot like school. At RCA, however, it was a different story. "They had nothing but labs, filled with equipment. And my eyes lit up. Look at this—endless equipment to play with, what a dream come true!" Even though the salary was the lowest offered, Maruska signed up with RCA on the spot.

A brick and steel-frame complex located in a leafy suburb of Princeton, not far from the university, RCA's Sarnoff Research Center had long been hog heaven for researchers. In the mid-1950s, Herbert Kroemer found it a wonderful place where he worked surrounded by talented people. "[RCA founder David] Sarnoff's charter for RCA labs was very broad," Kroemer told me. "He was not concerned about doing something that would be immediately useful within two or three years. He was taking a long-range perspective. We had pretty much complete freedom to work on whatever we saw fit. I worked on some pretty crazy stuff."

By the time Maruska arrived in 1966, this open-ended ethos remained pretty much unchanged. "The place was just so much fun to work at," he recalled. "I've never had fun at a place like I had there. Everybody was very enthused . . . it was all at the cutting edge, things that no one had done before were being done, people were scrambling around in the halls with just endless excitement. I would come in on the evenings, on weekends. No one told me to. The thing was just so exciting. Nobody bothered you with stuff, if there was some new idea and you wanted to work on something, then it was—Sure!"

The Sarnoff Center was, quite literally, an inventions factory. The many electronic technologies that originated there include the liquid crystal display and the technology that enables color LCDs, the thin-film transistor; complementary metal oxide semiconductor technology (the microchip industry's mainstream process); and the amorphous silicon solar cell.

Though nominally sponsored by RCA's computer lab, Maruska wound up being taken under the wing of Jacques Pankove, a kindly senior researcher. Pankove was born Jakob Pantchechnikoff in Russia in 1922. His Jewish family fled that year's pogroms to France, where six-month-old Jakob became Jacques. In 1942 Jacques was forced to flee again, the Nazis this time. Luckily for him, in Marseilles he managed to jump on board a ship bound for America. He ended up studying at the University of California at Berkeley. Upon marrying, to avoid saddling his wife and children with such a cumbersome handle as Pantchechnikoff, he shortened the name to Pankove.

In 1966 Pankove was working on Holonyak-style gallium arsenide phosphide LEDs and lasers. It was a wonderful way for young Maruska to get his feet wet. "Even though I had never done any real research before, I got involved with a project that ended up with me writing a journal paper. Which is pretty exciting, to start off from nowhere."

The only other electronics R&D facility that could compare to the Sarnoff Center was Bell Laboratories, up the road from Princeton in

Murray Hill, New Jersey. But whereas Bell Labs' focus was telecommunications, RCA's forte was television. In this field, the biggest research target—the Holy Grail—was to invent a replacement for the bulky cathode ray tube. This would be a flat-screen television that could be hung on the wall like a painting.

In 1964 a group at the labs had come up with one possibility, the liquid crystal display. But in their early days LCDs were only capable of displaying black and white. What they really needed was some sort of device that was capable of emitting the primary colors, red, green, and blue. Holonyak's gallium arsenide phosphide alloy produced red. Bell Labs researchers had recently succeeded at getting gallium phosphide to produce green. That left blue. It was thus entirely natural that Maruska's boss, Jim Tietjen, should come into the lab where the young man was growing some crystals on that mid-May morning and announce: "I've got a great idea—I think we can make a flat panel display. All we need is a blue LED." As the head of the materials research group, Tietjen had the funds at his disposal to get things going. He even had a hunch about what material to tap: gallium nitride. The periodic table suggested that GaN ought to have the right sort of bandgap to do blue. Now the question was, how to grow it?

Most of the old papers on gallium nitride, dating back to the 1930s, were in German. Maruska, who came from a German-speaking family, was able to read them. Gallium nitride had first been synthesized in powder form. As yet, however, no one had tried to grow films of single-crystal GaN, the kind of material you would need to make an LED. This is what Maruska set out to do. He went out, ordered a big canister of ammonia—atmospheric nitrogen is too chemically stable—installed it, got some gallium, and set to work.

Sapphire was chosen as the substrate on which to grow films. Unlike most materials it did not react with ammonia, hence it would not rot. Since RCA was running a large program fabricating radiation-resistant silicon-on-sapphire devices for aerospace applications, plenty of sapphire was available at the labs. For the first nine months or so, owing to a misinterpretation of the early German work, Maruska kept the growth temperature of his reactor way too low. The resultant material was polycrystalline gunk that you could wipe off the substrate with your finger.

"The nice thing was, nobody came and said, Hey, do something useful! They let the time slide by. It was, Keep trying, even though what you're doing is not working, try something else. Somehow I don't think you're allowed to do that anymore, but that's what made the atmosphere at the Sarnoff Center wonderful."

The first breakthrough came on November 22, 1968. Maruska was sitting at his desk wondering in desperation what he should try next. After nine futile months his whole program had turned out a total failure. What if they fired him? He would be drafted and sent off to fight—and possibly die—in Vietnam. Then an idea popped into his head. "I said, What the hell—why don't I make believe I'm growing gallium arsenide?" That meant cranking up the temperature of his furnace by several hundred degrees. At the end of the run, he took out the fingernail-sized piece of sapphire he had been using as his substrate. It was clear, with a mirror-like surface.

"I said, Nyah, nothing grew. But then I took it to the balance and weighed it, and son of a gun! There was a huge increase in weight." Seeking immediate confirmation, he grabbed the sample and ran down to the basement where the analytical center was located. There they took an x-ray diffraction photo. Sure enough, this confirmed that Maruska had made the first-ever single-crystal gallium nitride. For a young man without a PhD, it was a considerable achievement.

Work on gallium nitride at RCA was done under contract for the Department of Defense. The military was interested in flat panel displays for use in its mobile command and control centers. That meant Maruska could get deferrals from being drafted for Vietnam. But in 1970 he would turn twenty-six: too old for the draft. Now he could go back to grad school to get his PhD. This, as Shuji Nakamura would discover at the University of Florida some twenty years later, was de rigueur for anybody who did not want to be seen as second-rate. "Unless you had gone through the process of getting a PhD you were not treated as a full-fledged researcher; for example, you weren't included at staff meetings, you were just considered a technician."

Happily, Tietjen was able to arrange a corporate fellowship for Maruska to underwrite his studies. RCA would continue to pay his salary.

Considering that Maruska's course would take more than three years, this was exceptionally generous. There were two caveats. One was the company stipulated that his thesis topic had to be, Make a blue LED (and don't come back until you have one). No arguments there, that was what he wanted to do anyway. The other was the company would pick the school where Maruska went to do his thesis work. He had wanted to go to the University of Southern California. That made sense because a professor called Murray Gershenzon, who had been the first scientist to work on gallium phosphide at Bell Labs, had recently started a research program on gallium nitride there. But RCA's management refused: Southern Cal was a *football* school. Maruska would go to Stanford, a much more prestigious seat of learning. (Ironically, the year after he arrived, Stanford would win the Rose Bowl.)

In January 1970, some six months before Maruska's departure for Palo Alto, his mentor Jacques Pankove returned from a year's sabbatical at his alma mater, UC Berkeley. Pankove asked his protégé what he had been doing in his absence. When Maruska told him about this new direct-bandgap semiconductor, gallium nitride, Pankove became excited. Especially after he measured the photoluminescence of one of Maruska's samples, a sure indicator of a material's quality. Pankove got such a tremendous signal that it pinned the needle on the instrument's dial. At that point, Pankove decided that he, too, would work on GaN.

Although they were working in different labs at opposite ends of the building, a close collaboration between the two quickly blossomed. "I wound up growing the devices, then I'd carry them down the hall to Jacques. He had all this analytical equipment, he'd make all the measurements, then he'd come back and say, Why don't you try this, grow one this thick, or that thick?" Pankove became the young man's behind-the-scenes thesis adviser. The two men became, and would remain, close friends.

When Maruska headed west to Stanford that summer, Ed Miller, a research chemist at the center, took over as Pankove's collaborator. Miller grew films, passing them along to Pankove to fabricate and evaluate devices. In the summer of 1971 came exciting news. Pankove had observed for the first time electroluminescence in gallium nitride. They attached wires to a sample of GaN, and zapped it with a large current. The sample glowed with blue-violet light. Encouraged by this result, they pushed on urgently. "The period of time in 1971 and 1972 gave me what everyone who is involved with basic research would dream of," Miller recalled in a memoir. "That is, a hectic, exciting, productive effort to advance the frontier of science."

Pankove and Miller made the world's first gallium nitride LED. This was not your regular, positive-negative–style junction device. Despite their efforts, it proved impossible to produce positive-type material. The GaN films grown thus far were all intrinsically negative-type conducting material. There was, however, an alternative way to make light emitting diodes. They built a three-layer sandwich consisting of negative-type gallium nitride, a layer of insulating gallium nitride containing zinc (a hole-donating dopant used to make p-type gallium arsenide), with transparent indium metal on top.

Known as metal-insulator-semiconductor, or MIS, this type of LED works using a high electrical field to excite electrons. As these electrons careen through the crystal, they smash into the p-type dopant atoms. The collisions knock some electrons up into the conduction band. The upheaval is momentary, the knocked-up electrons quickly returning whence they came, giving off light in the process. But the chance of such collisions occurring is small, thus the efficiency is low and cannot be much improved. Nonetheless, MIS LEDs do emit some light. The first one Pankove and Miller made shone green. Other colors would soon follow.

Meanwhile, over on the West Coast, Maruska's mission at Stanford was to build a blue LED. Aware that a PhD thesis has to be based on original research, he decided to adopt a slightly different approach from his colleagues back east. He would substitute magnesium for zinc as the hole donor, a material that required less energy to activate. He built a reactor identical to the one back in Princeton. By June 1972 it was up and running.

In those early days, the equipment was primitive. Maruska's reactor was a horizontal quartz tube about an inch in diameter. In it you placed a little quartz boat filled with liquid gallium, then closed the tube. You loaded your dopant into a little quartz bucket that was attached by a hook to the end of a quartz rod. You lowered this rod via a side tube into the main reactor tube. There, heated by the furnace, the dopant would evaporate. You didn't know how much material was left in the bucket at any given moment, or exactly where it was, or what the temperature was. If you did get a result, it was not so much good science as good luck.

Maruska got himself some little balls of magnesium and loaded them in the quartz bucket. He lowered it gingerly into the reactor, then—disaster! Unbeknownst to him, liquid magnesium dissolves quartz. It dissolved the bucket, fell into the tube, then dissolved that, too. Finally, adding to his woes, the hydrogen carrier gas burst into flames. "So there's my thesis: up in smoke!"

Happily, Maruska managed to solve the problem by substituting graphite for quartz. Soon he had produced films doped with magnesium. He sent samples back to Pankove, who reported bright violet photoluminescence, thus raising everyone's hopes. Seeking to observe electroluminescence, Maruska attached electrodes to a sample and applied a high voltage. The first results were not impressive. "I had to turn out the lights in the lab, and sit there for fifteen or twenty minutes till my eyes adapted to the dark. Convinced that the film had lit up, I went and got a camera, put it in front of the material, opened the shutter, and left it open for about an hour. Then I came back, developed the film, and sure enough—there it was! Which is pretty hilarious, but that's how inventions are made."

Not long afterward, on July 7, 1972, having made various changes to the growing conditions, Maruska fabricated an MIS LED that was bright enough to see in a well-lit room. It emitted violet rather than blue light. Magnesium doping of gallium nitride would subsequently become the basis for all bright blue light emitters.

Back in Princeton, Pankove and Miller had managed to produce blue LEDs. They were not very bright—gallium nitride MIS LEDs would never be more than 0.01 percent efficient at converting electricity into light—but at least they worked. Otherwise, the news from RCA was all bad. In 1971 David Sarnoff died. His son Bobby had long since taken over day-to-day running of the company. The younger Sarnoff was not a good manager: he succumbed to the fad for corporate diversification, among other things, acquiring a carpet maker, a poultry farm, and a car rental company. The joke was that RCA no longer stood for Radio Corporation of America, but for "Rugs, Chickens, and Autos."

Also in 1971, RCA pulled out of the computer business, taking a massive $250 million write-off in the process. In November that year, Ed Miller wrote a letter to Herb Maruska, detailing the firings and layoffs that had recently been announced at the labs as part of a 10 percent cut in personnel and budget. On a visit to Princeton around this time, Maruska noted that "everyone was looking at the floor. All the fun had gone."

The gallium nitride group at RCA envied and feared their counterparts at Bell Labs. Envied, because there were only three of them in the group at the Sarnoff Center, whereas the word was that at Murray Hill about ten staffers had been assigned to GaN R&D. Feared, because the phone company's researchers might beat them to their elusive goal. Then, one evening at a conference in New Hampshire in the summer of 1973, Miller happened to be sitting next to the head of the materials research group at Bell Labs. He grabbed the chance to ask how the gallium nitride work at Murray Hill was going. Miller was astonished by his answer. Unable to make p-type GaN, they were shutting down their program. The RCA man was further dismayed to note that also at the table was Dave Richmond, the head of the materials research group from the Sarnoff Center. His boss had heard every word that was said.

For the next few months, however, the signs remained favorable. Richmond began calling Maruska, who had now been three years at Stanford, asking him when he would finish his PhD thesis. He reminded Maruska of his promise to return to Princeton, threatening to fire him if he didn't. Tietjen, recently promoted to director of the materials research laboratory and excited about the prospects for gallium nitride, was apparently planning to hire more people in order to expand the program. It was thus in an optimistic frame of mind that Maruska returned to New Jersey, bringing with him one violet LED. He showed up at the Sarnoff Center on the first working day of the new year. "I came in through the door, with my thesis under my arm, and a big smile from ear-to-ear, saying, Look, I arrived when you told me to, with what you told me to do!" What he was about to hear soon wiped the smile off his face.

Tietjen summoned Pankove, Maruska, and Miller into his office. Maruska remembered the scene vividly. "Tietjen was an extremely neat, well-organized man. There was not a paper or pen on his desk, not even a telephone." The director looked at the trio across the clear polished desktop and, with deep sadness in his voice, told them that he was canceling gallium nitride research, a program that he himself had initiated. His budget had been cut drastically. Tietjen could no longer afford to continue with research that showed no sign of leading to a commercial product in the short term. "You guys have bled me dry with nothing useful to show for it. I have no more blood to give—I'm sorry, but I can't support you anymore."

For Maruska in particular, the news that his beloved program, to which he had devoted more than five years of his life, had been canceled was a devastating blow. "It was like getting smacked in the head with a two-by-four," he told me sorrowfully. Thirty years later, he would still have nightmares about it.

Not one to give up without a fight, Maruska quickly set about building one of the world's first MOCVD reactors. Had he and Pankove been given the chance to experiment with it, RCA might have produced a bright blue LED well before the rest of the world. However, as 1974 came to an end, his new reactor wound up in a dumpster and he himself wound up out on Highway 1, jobless.

Pankove soldiered on for a little longer, devising alphanumeric displays, trying to generate interest in blue LED technology from RCA's marketing people. As at Monsanto when George Craford invented a yellow LED, however, the response from RCA was, What we need is not a different color but a cheaper LED. Since using sapphire substrates meant that blue LEDs would always be more expensive than red ones, it was determined that there would be no market for them.

Eventually, with no commercial interest in the offing, Pankove was ordered to quit. "I was told, Every time the vice president of the laboratories sees gallium nitride in the progress reports, his face turns red and he goes, What—are you still working on this?" Pankove stopped mentioning GaN in his progress reports. He continued the work under the table. But by early 1974, gallium nitride research at RCA was effectively over.

*   *   *

It seems remarkable, with hindsight, that a research topic that had been deemed not worth proceeding with by the top two electronics laboratories in the world should immediately have been picked up by a researcher at a Japanese laboratory. Even more remarkable was that the laboratory in question should belong to Matsushita, the most pragmatic and market-driven of Japanese consumer electronics companies. But research laboratories sometimes function as a kind of corporate status symbol. Matsushita already had a central research laboratory in Osaka, where the company was headquartered. In the early 1960s the company founded another laboratory. This was located in Tokyo, three hours away from Osaka by train. Its focus would supposedly be basic—that is, nonproduct-related—research.

Few of Matsushita's top executives had had more than a high school education. Konosuke Matsushita, the company's founder, had (like David Sarnoff) almost no formal schooling. It pleased the old man to be able to hire highly educated white coats from prestigious national universities to staff his new laboratory. But as quid pro quo for joining Matsushita, the scientists insisted on being allowed to do whatever they liked, regardless of commercial applicability. In the case of Isamu Akasaki, who joined Matsushita from Nagoya University in 1964, that meant working on compound semiconductors like gallium arsenide.

Akasaki was born in 1929, which makes him an almost exact contemporary of Nick Holonyak (b. 1928), the inventor of the LED. By the late sixties, Akasaki was fabricating Holonyak-style gallium arsenide phosphide red LEDs. From there, he proceeded to make Bell Labs–style gallium phosphide green LEDs. Having succeeded in that, too, he was naturally eager to move on to the next thing, i.e., blue. In 1973 Akasaki began work on gallium nitride. The following year, he grew his first film of the material. Encouraged by this result, he did what any self-respecting Japanese industrial researcher of the period would have done: he went straight to the Ministry of International Trade and Industry (MITI) and asked for some money.

As it happened, Izuo Hayashi, the semiconductor laser pioneer who had recently returned from Bell Labs to work for NEC, had just put in a similar proposal. The result was a combined Japanese government-funded consortium to develop blue light emitters. It ran for three years, from 1975 to 1978. Matsushita put the resultant blue MIS-type LEDs into trial production. A few thousand devices were fabricated—some were eventually sold as samples—but mass production was not possible. There were far too many cracks in the crystal. And without p-type material, the conversion efficiency of electricity to light was never going to reach marketable levels.

Akasaki was not downhearted by this failure. On the contrary: "I had developed a gut feeling that the goal was within my grasp. After that, I expanded my research activities around this point, just like driving a crack into a wedge."

Seeking a way of growing higher quality, thinner films, Akasaki switched from vapor phase epitaxy, the method Maruska had used, to the new method of metal organic chemical vapor deposition. In 1981 he left Matsushita and returned to Nagoya University as a full professor. It was like starting from scratch. His laboratory was almost empty. It had very little equipment and no clean room facilities. Worse, it proved impossible to persuade the Japanese Ministry of Education to fund research on gallium nitride. By the early 1980s the field was almost moribund, with virtually no papers being published. His grant proposals routinely being rejected, Akasaki was forced to resort to subterfuge. He diverted money earmarked for work on conventional compound semiconductors into nitride research.

For the first ten years, Akasaki had done most of the hands-on crystal growing himself. Now, with a university research lab to run, this was no longer possible. He hired Hiroshi Amano, one of his young graduate students, as his assistant. Years later Amano would explain what had attracted him about the theme of the blue light emitting diode.

"The most important [reason] was that, at that time, nobody was able to succeed . . . [so] there was the chance to become the top runner. And the theme stood out because of its closeness to the creation of a final product for people [to use]."

Amano began his research knowing next to nothing about crystal growth. Naturally, he had many bitter learning experiences along the way. In attempting to produce high-quality films of gallium nitride, his first challenge was dealing with parasitic reactions in the growth chamber that caused white powder to form and ruin the material. This, as we shall see in the next chapter, is a problem that some years later Nakamura would also confront.

Like Nakamura, Amano dealt with it by modifying the way the gases flowed in the reactor. Like Nakamura, lack of budget plus desire to proceed rapidly meant that Amano had to roll up his sleeves and make the parts himself. "At first, they were failures. After thirty or forty attempts I made the glass workings the way I wanted them, which resulted in the desired gas flow."

Like Nakamura, Amano worked incessantly. "Not a day had I been absent. Counting up the number of times of experiments, I had done over one thousand five hundred experiments. But the obtained results were always like frosted glass. I had been working body and soul on these experiments, but in vain. During this period, the other master's course students . . . got jobs lined up, but I had no hope of getting a job and necessarily decided to stay on as a PhD student. [But] the results of the experiments didn't amount to much, so I could not prepare the scenario for my master's thesis at all. It was a miserable situation for me."

Although similarly determined, Amano lacked Nakamura's hard-won self-confidence. ("I was perhaps the worst student in the Akasaki laboratory.") Happily, he was not working on his own. He could always go to his professor for guidance about what to do next. Like a good mentor, Akasaki didn't explain everything in detail. He just provided a clear direction, leaving it up to his student to make his own decisions.

Then one day in 1986, as so often in science, serendipity intervened. Something was wrong with the heater in the MOCVD reactor. An idea popped into Amano's head. Instead of gallium nitride, he grew a thin layer of aluminum nitride. This softer material could be deposited at a lower temperature. Akasaki had worked extensively on aluminum nitride at Matsushita. By the time Amano had grown this film, the glitch in the heater had sorted itself out. He was able to raise the temperature and grow another film, this time of gallium nitride.

When he took his sample out, the sapphire looked as if there was nothing on it. Amano wondered whether he had forgotten to turn on the gases. But when he checked the surface with a microscope, like Maruska with his first-ever sample of single-crystal gallium nitride, he saw that a film had grown. It was an unforgettable moment: the excitement made his heart race. On top of his buffer layer of aluminum nitride, instead of frosted glass, he had grown a mirror-smooth layer of high-quality gallium nitride.

That same year Akasaki embarked upon a collaboration to commercialize the technology for making MIS-type blue LEDs. It was based on patents his group had filed that were now owned by Nagoya University. The chosen partner was Toyoda Gosei, a local firm that was a leading supplier to the carmaker Toyota. Toyoda Gosei's stock-in-trade was molded rubber and plastic auto components such as brake hoses, fenders, and steering wheels. Like most large Japanese firms, however, the company was always on the lookout for opportunities to diversify.

Around that time Stanley Electric, another Toyota supplier, was beginning to make headlines with its high-brightness red LED brake lights. No doubt Toyoda Gosei hired some of Akasaki's graduates with a view to emulating Stanley's success. At any rate, the collaboration commenced. As with Stanley, funding came from the Japan Science and Technology Corporation, a government agency tasked with transferring technology from national universities to the private sector.

Louis Pasteur said that chance favors the prepared mind. So it was with Amano. In 1989, after another three years of hard slog, he finally made the crucial breakthrough, the one that had stymied the American researchers of the previous generation. It was the result of an accidental discovery. Amano was examining a sample of zinc-doped gallium nitride in a scanning electron microscope. As the beam of electrons scanned over the material, he noticed "a very curious phenomenon"; namely, the sample was glowing. As time went by, the luminescence increased. But the material was still not positive-type. On a hunch, Amano switched from zinc as his hole-donating dopant to magnesium, the same material that Maruska had pioneered. He irradiated the resultant sample in the electron microscope. Immediately, he was able to produce p-type gallium nitride.

It was a result that, many years later, would cause Maruska to groan with frustration as he contemplated what might have been. After all, he too had spent countless hours examining magnesium-doped samples of gallium nitride under the scanning electron microscope. Why had none of them ever emitted so much as a glimmer of light? The reason, it turns out, is that all of the samples Maruska prepared back in the early days were inadvertently contaminated with oxygen, which mops up any available holes in the material. By the time Akasaki and Amano were performing their experiments, fifteen years later, they had switched to more advanced gas transport systems that did not involve oxygen and that used reactants that were purer, containing no residual moisture.

Akasaki and Amano announced their wonderful discovery at a conference in Japan. After the presentation, they took questions from the audience. One question came from someone called Nakamura, a researcher they had never previously encountered, who said he worked for a company called Nichia. They did not realize it then, but the race to build a bright blue LED was on.

The world's first proper—that is to say, containing a *p-n junction*—gallium nitride LED made its debut at a conference in Los Angeles in 1989. But Akasaki and Amano's device was not very bright, and the light it emitted was as much violet as blue. The human eye is not very sensitive to violet light; blue light is much easier to see. The next step was to build a double-heterostructure device, which would shine brighter. This took the pair another two years. Akasaki reported their new results at the fall meeting of the Materials Research Society in Boston in December 1991. With patents pending, however, he was unwilling to risk demonstrating the new LED—at least not in public. As it happened, Maruska was also attending the meeting. A mutual friend introduced him to the Japanese professor. Maruska, who had long since quit the nitrides field, did not know who Akasaki was. Akasaki, of course, was well aware of the name Maruska.

"Then Akasaki pulls me aside, and it's funny because he's about half as big as I am, and he whispers in my ear: I could show you something if you come to my hotel room at nine o'clock tonight. He gave me the room number, and I went and knocked on his door. The door opens just a crack, he sticks his hand through it, holds this blue LED in front of my eyes, turns it on, and says, Look at that. And I shouted, Jesus!—it's so bright! I almost fell over backwards into the hallway. He said, That's all for now, then shut the door. I said, Oh my God—the problem's been solved!"

Not quite. Though the output from the new Nagoya University LED was ten times higher than that of the silicon carbide devices that a new American start-up called Cree had recently begun selling, the light was still not exactly what you would call bright. Unless of course it was shined right into your eyes from close up against a dark background, as Akasaki had done to Maruska.

Now, coming down the home stretch, Akasaki and Amano ran out of puff. In 1992 Akasaki turned sixty-three. That year he would be obliged to take compulsory retirement from Nagoya, a national university, and relocate his lab to Meijo University, a nearby private school. The move would of necessity cause a major disruption to his research efforts. Meanwhile, as we shall see, Amano would undergo a crisis of confidence and falter at the last hurdle. This time, there would be no sage advice from Akasaki to help him over. And all the while Nakamura was coming up unseen behind them, gaining fast.

By 1992, as the next chapter describes, Shuji had already overtaken his rivals. They had transferred the technology for making double-heterostructure LEDs to their commercial partner, Toyoda Gosei. In a corporate brochure dated April 1993 the company boasted that it "currently heads the world in research and development of blue LEDs." If it did, it was not for long.

At the end of May, the US Air Force sponsored a study group to Japan that visited both Nakamura and Akasaki. Knowing that the Americans had been to see Nakamura, Akasaki was eager to know about the younger man's progress in growing indium gallium nitride. He did not have to wait long to find out. In November that year, about a week before Nichia made its astonishing announcement, Akasaki got a courtesy phone call from Nobuo Ogawa telling him the bad news.

The race to build a bright blue LED was over: Nakamura had won.

* * *

Fast-forward four years to December 1997, to another fall meeting of the Materials Research Society in Boston. A banquet was held at the John F. Kennedy Library to honor Jacques Pankove. There, the title "godfather of gallium nitride" was conferred on him. Sitting next to Pankove at the festive meal was the "father of gallium nitride," Herb Maruska. Master of ceremonies that night was Isamu Akasaki, who might perhaps be dubbed the "stepfather of gallium nitride." Attendees were presented with a bright blue Nichia LED light pen, which they raised in a toast to the GaN pioneers.

By 1997 the first wave of gallium nitride products was becoming visible to the general public, most notably, in the form of green LED traffic lights. The first time Maruska saw the lights, he was so amazed, he just stood there and stared. It was almost thirty years since he had begun his gallium nitride research by photocopying those pages at the Sarnoff Center. "All this time I had been imagining that gallium nitride LEDs were going to be successful one day, then I'm standing there saying, Look—they're real! I took a picture of the lights, and obviously I had a grin from ear to ear, it was just such a joy to see them."

CHAPTER 5

The written instructions to Shuji Nakamura were crystal clear. They said: "Stop work on gallium nitride immediately." The instructions were signed by Nichia's new president, Eiji Ogawa. During Nakamura's year in the United States, a changing of the guard had taken place at Nichia. Nobuo Ogawa, then aged seventy-seven, had handed over the reins to his son-in-law, and the old man became chairman.*

If Nobuo was an entrepreneur, then Eiji was a manager. "He doesn't come across as a tough businessman," commented one Nichia customer who dealt directly with the younger Ogawa, "but he has a mind of steel, and he remembers everything—he knows what his figures are, what his factories cost to run, and what his cost of development is."

Eiji was a very different sort of character than Nobuo. The old man was voluble, a good speechmaker and raconteur, able to chat casually with employees in a friendly way. The new president lacked such interpersonal skills. Eiji found it difficult to talk to people. Nakamura recalled that he would creep up behind you—you couldn't hear his footsteps—then simply stare at you.

---

*Eiji had taken the Ogawa family name on marriage, as Nobuo himself had done, a common Japanese practice to maintain the name of the dominant clan. Nobuo did have two sons of his own. Both had worked at Nichia—one joined the same year as Nakamura—but both subsequently left.

If Nobuo caught his employees slacking, dozing at their desks, or sneaking a smoke, he would shout: "What do you think you're doing?! We'll go bankrupt if you behave like that!" Then the worker would apologize, and that would be the end of the matter. But if Eiji caught someone napping or puffing on a cigarette or whatever, he wouldn't tell them off on the spot. The president would go back to his office, write down the infraction, then give the name of the offending employee to his immediate boss, leaving it to the latter to administer an admonition. Writing notes, not verbal communication, was Eiji's style. Hence the instructions to Nakamura to cease what he was doing.

The crisis had arisen suddenly in the wake of a visit to Nichia by a VIP. This man was a senior R&D manager at Matsushita, by far Nichia's biggest customer. As it happened, the VIP's speciality was light emitting diodes and semiconductor lasers. In fact, it had been largely his suggestions that had led to Nakamura's previous forays into the LED field, the ones that had turned out commercial debacles. Nichia's salesmen were always eager for this top executive to come down to Tokushima, where they would entertain him like royalty.

Wishing to flatter their esteemed customer, and knowing that the Matsushita man liked the sound of his own voice—he was forever pontificating about something or other—the salesmen would typically invite him to give a talk. The title of his lecture that year, 1990, was by coincidence "Research on Blue LEDs." He talked about how Matsushita itself was working on blue LEDs, basing its research on the material zinc selenide. Although progress was slow, it was clear that zinc selenide was the best way forward. Then he proceeded to dismiss other candidate materials, emphasizing that "gallium nitride has no future." Nakamura was shocked to hear this. Thus far he had been able to get away with doing as he liked, because no one else at Nichia knew anything about semiconductors. Now, he realized, he was in serious trouble.

Worse was to come. Following the lecture, without warning the visitor arrived in Shuji's lab, escorted by Eiji. This was an unwelcome surprise. Nichia was normally the most secretive of companies. Nakamura's research on the bright blue LED was supposed to be highly confidential.

Now a horrified Shuji was confronted by this senior researcher whom he knew well from a rival firm. Recognizing the MOCVD system, the visitor asked Nakamura what he was working on. Shuji couldn't very well tell him the truth, especially not after hearing the Matsushita man's negative views on gallium nitride. So he lied, saying infrared gallium arsenide LEDs. Ever ready with a handy hint, the visitor kindly suggested that Nakamura should use his system to grow high-speed gallium arsenide transistors. There would, he predicted, be a great future for such devices in the cell phone market.

Back at the president's office, the VIP repeated his suggestion for Eiji's benefit. After his departure, Nichia's president scribbled instructions to Nakamura: "Stop work on gallium nitride immediately and start development of high-speed transistors." Eiji had been against Nakamura's plans from the outset, seeing them as just a further waste of Nichia's money. By that point, however, Shuji had already determined that he would ignore all instructions from the company. So he simply crumpled up the note and tossed it in the bin. Over the next few months, a succession of similar written orders from the president telling him to desist would be similarly disposed of. Nothing was going to stop Nakamura now.

✳ ✳ ✳

In *What Mad Pursuit*, his book of musings on the nature of scientific discovery, Francis Crick notes "the curious attitude of scientists working on 'hopeless' subjects. Contrary to what one might at first expect, they are all buoyed up by irrepressible optimism. I believe there is a simple explanation for this. Anyone without such optimism simply leaves the field and takes up some other line of work. Only the optimists remain."

This was certainly true of Nakamura as he set out on his quest to develop a bright blue LED using gallium nitride as his base material. "People often say I'm optimistic. But if I were pessimistic I wouldn't be able to go on. Unless you feel sure that you can succeed, you cannot continue diligently doing experiments every day."

The life of a researcher is not the series of eureka moments that the popular press might lead us to believe. On the contrary, it is for the most part dull and monotonous. Especially so in Shuji's case, because he had no colleagues with whom he could discuss his work. His solitary routine seldom varied. He avoided distractions like socializing with his work-mates. Other than New Year's Day, when everything in Japan shuts down, he never took time off. He would get into work around seven every morning, leaving around seven in the evening. He would go home, eat dinner with his family, have a bath, and go to bed around eleven. All the while he would be musing about his work. Relaxing at home helped him to concentrate. Preoccupied with a problem, he would withdraw into himself. In this meditative state, he would not hear his wife when she called. Sometimes, walking down the street, staring straight ahead with unseeing eyes, he would bump into utility poles.

"When I start something, I tend to get obsessed by it. I think about it all the time, sleeping or awake. And I am very impatient. When I have an idea I must try it immediately. If I get ten ideas, I try all ten; if I get twenty, I try twenty." Every morning, acting on the ideas he had had the previous evening, Nakamura would make some modifications to his equipment. Every afternoon, he would grow some samples. This pattern continued for three-and-a-half years, seven days a week, until Shuji had achieved what he set out to do. It was a one-man product development marathon.

Modifying the equipment was the key to his success. It laid the foundation for growing high-quality thin films of gallium nitride crystal, the sine qua non of bright blue LEDs. For the first three months after he began his experiments, Shuji tried making minor adjustments to the machine. It was frustrating work. The equipment had not been designed to grow gallium nitride. The wafer heater got corroded by ammonia gas and would often cut out. Nakamura eventually concluded that he was going to have to make major changes to the system. Once again he would have to become a tradesman, or rather, tradesmen: plumber, welder, electrician—whatever it took. He rolled up his sleeves, took the equipment apart, then put it back together exactly the way he wanted it.

Rooting around the innards of an MOCVD system is not for the faint-hearted. It takes a variety of skills. According to Bob Karlicek, a former

Bell Labs chemist who has spent much of his career working with MOCVD systems, "You have to be able to do fluid dynamics, you have to know the gas-phase chemistry, what materials are compatible with the gases at high temperatures, and there's also a lot of mechanical design that goes into building reactors. Only a very few people have the knowledge—or the courage—to go inside a reactor and really optimize it."

Nakamura was one of the few. He bent the steel pipe, changing the height and the angle at which it was attached to the reaction chamber. He welded quartz tube, cut high-purity carbon, redid the wiring. He even altered the shape of the nozzles where the gas came out.

Elite researchers at big firms prefer not to dirty their hands monkeying with the plumbing: that is what technicians are paid for. If at all possible, most MOCVD researchers would rather not modify their equipment. When modification is unavoidable, they often have to ask the manufacturer to do it for them. That typically means having to wait for several months before they can try out a new idea.

The ability to remodel his reactor himself thus gave Nakamura a huge competitive advantage. There was nothing stopping him; he could work as fast as he wanted. His motto was: Remodel in the morning, experiment in the afternoon. If he didn't finish remodeling in the morning, then he couldn't do experiments in the afternoon. Such urgency was not because Shuji was worried that other researchers might overtake him. That seemed unlikely. To his knowledge only one other group—that of Professor Akasaki at Nagoya University—was working on gallium nitride LEDs. And fiercely competitive as ever, Shuji was confident that, when a university professor and a company compete, the company usually wins. It was because, impatient by nature, Nakamura was eager to see the results of the changes he had made.

In order to grow high-quality thin films of gallium nitride it was necessary to overcome one major problem. You pump gases into the reactor jar. There, they are supposed to combine with each other on the surface of the wafer. But nitrides are notoriously vulnerable to parasitic reactions. That is, if the density of their molecules becomes too high, the gases react with each other spontaneously to form what is known in chemical jargon as an adduct—in this case, a white powder that researchers call "snow."

Flakes of snow fall on the wafer, ruining the film. Much ingenuity therefore goes into designing reactors so that the gases are injected separately,
keeping them apart as they flow down to the wafer.

Nakamura hit on a novel way of doing this. He conceived an entirely
new design. Conventional MOCVD reactors have a single inlet at the top
of the jar. Gases are injected through a water-cooled flange. The huge difference in temperature between the entrance flange and the heated wafer
necessitates that the two be kept some distance apart. In this open space,
the precursor gases are free to react with each other. Nakamura realized
that the source of the parasitic reaction problem was convection caused
by heat rising from the wafer, producing a buoyancy effect that made the
hot gases rise and circulate. Convection made growth of ultra-thin films
virtually impossible.

To solve the problem, he came up with the idea of adding a second
inlet located directly to the side of the wafer. Gases coming in from the
side would have no time for parasitic reactions. A heavy flow of inert
gas—tens of liters per second—injected perpendicular to the flow from
the side would suppress convection currents rising from the wafer. It
would keep the reactant gases pinned to the wafer surface, so that the
reaction would occur in the appointed place. He called this technique
"two-flow" MOCVD, an amazing achievement for someone who was not
by training a chemical engineer.

It was October 1990, a year and a half since he had begun his experiments. Looking back, Nakamura would come to see the development of
the two-flow system as his biggest breakthrough. Previously he had
served a ten-year self-taught apprenticeship in growing LEDs. Now he
had rebuilt a reactor with his own hands. This experience gave him an
intimate knowledge of the hardware that none of his rivals could match.
Almost immediately, Nakamura was able to grow better films of gallium
nitride than anyone had ever produced before.

Concerning his novel system, he wrote a patent application and a
paper. Shuji submitted the latter secretly, in defiance of Nichia's ban on
publications, to the *Japanese Journal of Applied Physics*. It was returned,
five or six times. On each occasion, the reviewer's comments concerned
not so much the contents of the paper as the citations at the end. Where

Nakamura had referred to an American author, the comment would read "That's incorrect—cite Akasaki's paper." Nakamura had naively imagined that the referee would be someone neutral. In fact, as he realized later, there was only one Japanese researcher in the nitrides field whom the journal could possibly have asked to review his paper. No prizes for guessing whom.

Eventually, Nakamura gave up, submitting the paper instead to the more prestigious American journal *Applied Physics Letters*. This time, the reviewer was more sympathetic. "I got a paper to review from *APL*," recalled Asif Khan, the leading US-based expert on MOCVD nitride growth. "It was this crazy guy suggesting a two-flow method. I had some of my guys look at it, and they said, This is not a good idea. But I said, No, guys, I like it, I think that this is something very new and innovative—it should be published." Nakamura's first paper appeared in *APL* in May 1991.

During the 1980s, Khan was more or less the only researcher in the United States working on the nitrides. Today, as we shall see in chapter 10, he heads a huge nitride laboratory at the University of South Carolina that, in terms of personnel and resources, is second only to the University of California at Santa Barbara, where Nakamura now works. At his palatial office in downtown Columbia, the state capital, I asked Khan what it takes, other than sheer persistence, to be a world-class crystal grower.

"One thing that's absolutely imperative is to have a very good background exposure to other fields, keeping an eye on what's going on, thinking about it, and seeing how to take advantage of it in your area. Second thing, no matter how good a crystal grower you are, if you don't have the right equipment you're not going anywhere. And the other thing I've seen is that, a lot of crystal growers tend to get very methodical: they never think out of the box. Nakamura is not that type; he's not the type of materials scientist who knows all the theory, all the fancy equations. He's the type that thinks out of the box, tries out something crazy, and sometimes things work out. I think he has been very lucky, but if he did not try this crazy stuff, then luck would not favor him."

Crazy stuff maybe, but there was also method in Nakamura's madness. To grow high-quality thin films of gallium nitride you need to get the recipe—temperature, timing, and other parameters—just right. Such optimization is not a matter of random choice and guesswork. It takes intuition guided by experience. For Nakamura, his intuition and experience were the most important factors. They kept him going through the dark and difficult days of late 1990.

By then Nichia had more or less given up on Nakamura. Initially, as usual, the company had repeatedly nagged him, demanding to know whether he had finished yet, how soon it would be before he could develop a product. But as time went by and no results emerged, his boss gradually stopped bugging him. Even Shuji's friends at the company who used to drop by his lab for a chat no longer bothered, such was his all-consuming preoccupation with his work. They left him alone to get on with it. In this extreme isolation, as failure followed failure, Nakamura became depressed. Oddly enough, however, the depression seemed not to be debilitating; rather, it had the opposite effect. The more depressed he was, the better his concentration became. Freed of distractions from the outside world, his mind was clear. He thought about how to make high-quality gallium nitride morning, noon, and night, every waking hour, even in his dreams. So deeply buried in his thoughts was he that, driving home one evening, he ran a red light and almost caused a nasty accident.

Then, at last, one fine winter's day, the clouds lifted. Everything was as usual: Nakamura arrived at work, put on his white coat, went into the clean room, and grew a thin film of gallium nitride crystal. He opened the reactor chamber and took out the transparent sapphire wafer on which the film had grown. The lunchtime siren sounded. Many employees went home for lunch, others brought to work lunches their wives had made for them. Shuji had his lunch delivered to the company. It came in a red plastic box. As he cut out a five-millimeter-square chunk from the wafer, he wondered idly to himself, What will be in my lunchbox today? Before eating lunch though, he had to finish his measurements.

He hooked up his tiny sample to the equipment used to measure electron mobility. In essence, mobility is a measure of the velocity at which electrons can zip through the crystal lattice. The better the quality of the crystal, the faster the electrons can whiz through it, the higher the measured mobility. Defects such as impurities, lattice dislocations, and micro-cracks impede motion. The printer spat out a number. It was surprisingly high: the world's best result up to that point, achieved by Akasaki and Amano at Nagoya University, was less than half as much. Shuji's heart leapt. Trying to stay calm, he checked to make sure he had not misread the printout. Then he confirmed that there was nothing wrong with the measuring equipment. Finally, to make sure that his sample was not abnormal, he cut out another chunk, and hooked it up. Same result: the evaluation was reproducible. Fantastic!

He had succeeded in making the best gallium nitride in the world. "It was the most exciting day of my life," he recalled. "I had never been number one in the world before. Now I was number one." He had attained pole position in his chosen field, a position he would maintain for the next nine years.

The electron mobility of his material was still nowhere near high enough to make a bright blue LED. But, Shuji reflected happily as he munched his way through the contents of his lunchbox, at least he could write a good paper about it. He also applied for a patent, figuring that when the company found out about the paper, they wouldn't be so angry if he had a patent, too. As it turned out, he was right. Having broken the world record for electron mobility, Nakamura now entered virgin territory. "There were no papers or reference materials. Even if I had wanted someone to teach me, I didn't know anyone who could. There were no numerical formulas written by learned professors, no wonderful equipment available in the marketplace."

But now the breakthroughs came thick and fast. Akasaki and Amano had blazed the trail, with their buffer layer (in 1986) and positive-type gallium nitride (in 1989). The buffer layer was necessary to mitigate the effect of the mismatch between the sapphire substrate and the gallium nitride layers deposited on it. By interposing a buffer you could grow smoother films. It was a technique that had previously been used in other fields—a good example of Khan's dictum about the need for broad awareness—but Akasaki and Amano were the first to apply it to the nitrides. For their buffer layer they had used aluminum nitride.

Nakamura was determined not to copy what the Nagoya University pair had achieved. He had made a strict rule for himself: research must be done independently, in a unique way. Copying was no good; besides, if he copied, there would inevitably be patent problems for his company down the track. So Nakamura used gallium nitride as the material for his buffer layer. He was able to achieve a smooth, mirrorlike surface that had better characteristics than aluminum nitride. The electron mobility of this new film was much higher than previously reported figures. It was March 1991: another world record, another paper secretly published.

But for a bright blue LED Shuji needed first to make positive-type gallium nitride. Akasaki and Amano had found that they could produce p-type GaN by irradiating the material with a low-intensity electron beam. This was a wonderful scientific discovery, but impractical from a technological point of view. E-beam equipment is expensive; e-beams have to be scanned laboriously across the material. The method was far too slow for application to any conceivable LED production line.

Akasaki and Amano announced their p-type discovery at a domestic Japanese conference in 1989, just after Shuji got back from Florida. Nakamura was in the audience. His initial reaction to their announcement was acute disappointment. He had wanted to be the first to achieve p-type GaN. At question time, Shuji put up his hand and asked what the hole concentration of their material was. Their answer told him that the material they had made was not very high quality. Realizing also that his rivals' technique for producing p-type material was not applicable outside of the lab, Shuji cheered up. At least they had demonstrated that it was possible to make p-type material. He would thus start his research on gallium nitride just as the hitherto most intractable problem in the field had been proved to be other than a showstopper. It was an incredible stroke of luck. Once again, his timing was impeccable.

Now, two years later, Nakamura repeated Akasaki and Amano's experiment to make p-type gallium nitride. By July 1991, armed with both negative- and positive-doped materials, Shuji was able to proceed rapidly to the next stage, actually making a simple LED. The research went quicker than expected: it took him just one month. He touched the probes to the surface of the wafer. The device lit up with a violet-blue light. It was not very bright, but it was 50 percent brighter than silicon carbide LEDs, which at that time were the only blues you could buy.

The outstanding question was longevity. Nakamura was worried that a fragile thin film shot through with ten billion defects per square centimeter could not continue to emit light for very long. In fact, he would be lucky if it lasted a few hours. But much to Nakamura's amazement, the little light stayed lit all afternoon. He went home that night, leaving it on. Next morning, with his heart thumping in his chest and a prayer on his lips, he came back into the lab to find that . . . it was *still* lit! He measured the output and was elated to discover that it had barely dropped from the previous day's value. But would the LED really last? It took him a few days to set up the equipment for a lifetime test. On the fifth day he performed the test. The night before, Shuji was so full of adrenaline he was not able to sleep. The lifetime turned out to be longer than one thousand hours.

Having confirmed the longevity, he went straight to see Nobuo Ogawa to tell him the good news. Pausing only to grab his camera, Nichia's chairman hurried back with Nakamura to his lab. But Ogawa's response when he saw the little light was not at all what Nakamura had hoped for. He just looked at it and said, "Ha! Is this such a big deal?" Nakamura turned out the lights in the lab so that Ogawa could see it better. But the old man just shook his head and said, "It's too dim—something as dim as that won't sell." Shuji sighed. He had been so happy to have made a technological breakthrough. Now he realized that in order to reach his goal of commercializing a bright blue LED, there was still a long way to go.

Shortly thereafter, bad news reached him from the United States. He read in a Japanese newspaper that a team working at the company 3M in St. Paul, Minnesota, had succeeded in operating a semiconductor laser made from zinc selenide. If they could make something as complex as a laser, he reasoned, then making an LED would be relatively straightforward. Shuji was devastated. Feeling that he had been defeated, he almost lost the will to continue his research.

But continue he did, focusing on the big problem of how to make high-quality p-type gallium nitride. Akasaki and Amano had not been able to figure out why e-beams caused the transformation. After giving the matter of the mechanism much thought, it occurred to Nakamura that perhaps it was merely heat that was causing the material to become p-type. In December 1991 he tried annealing magnesium-doped films, baking them in a nitrogen atmosphere oven. Sure enough, the resultant material was p-type.

It turned out that Maruska's hole-donating magnesium had been rendered electrically inactive by hydrogen, the main carrier gas used in the MOCVD process. To activate the magnesium, you could knock the hydrogen atoms out by bombarding them with e-beams, or you could simply boil them off with heat. Thermal annealing was simpler and much faster than the Nagoya pair's cumbersome e-beam method, and hence applicable to the production line. In addition, it also produced much better quality gallium nitride with a higher concentration of holes. E-beam penetration was very shallow, with only a very thin surface layer of the material becoming p-type. Thermal annealing converted the material to p-type all the way through. This was a major breakthrough.

$$* \quad * \quad *$$

A few months prior to this momentous discovery, a letter addressed to Shuji arrived unexpectedly from the United States. It was from Hadis Morkoc, a researcher at the University of Illinois, inviting him to speak at the first international workshop on nitrides. The workshop was to take place in St. Louis, Missouri, the following April. Through the papers he had been secretly publishing, Nakamura was beginning to make his presence in the world felt. Now the world wanted to know more.

Nakamura approached Eiji Ogawa to ask his permission to attend the workshop. The president immediately said no. It was not Nichia's policy to allow its researchers to attend conferences overseas. Nakamura faxed Morkoc his regrets. But the pushy professor would not take no for an answer. He faxed back saying, We need you at our conference. What can I do to get you to come? Nakamura replied saying there was no way, his president had vetoed the idea. Morkoc faxed back saying, I'll do anything. Nakamura suggested that he try contacting Eiji Ogawa and asking him directly.

Morkoc wrote Nichia's president a long letter explaining why it was vital that Nakamura should attend the workshop. Unexpectedly, Eiji relented: perhaps he could not refuse a request from such an eminent foreigner. Nakamura was made to understand, however, that this was not setting a precedent. It would be the last time that he would be allowed to attend an overseas conference. In actual fact, it would not be the last, but the first of many international conferences that Nakamura would attend. At most, Shuji would be the featured speaker.

The world's first nitrides conference was held at the Adam's Mark Hotel, just across from St. Louis's famous Gateway Arch, an appropriately iconic location for Nakamura's American coming-out party. There, for the first time, Nakamura met his peers in the nitrides field. There were perhaps forty of them, though few were really serious players. They turned out to be a heterogeneous bunch, a tribute to the ability of US universities to attract talent from all over the world: Asif Khan came from Pakistan, Hadis Morkoc from Turkey, Jacques Pankove from Russia via France, Ted Moustakas (of Boston University) from Greece. It seemed like almost the only American-born nitride researcher in attendance was Bob Davis of North Carolina State. And he was really more of a silicon carbide man.

As well as the scientists, also present were funding managers from various US defense research agencies. The air force in particular was interested in nitrides for use in missile warning systems. The navy wanted green lasers for underwater communications (water is transparent at green wavelengths). As so often happens in America, only the military has the cash to lavish on fostering exotic new technologies.

Oddly enough, the one leading researcher in the nitrides field attending the conference that Nakamura did not meet was his compatriot, Isamu Akasaki. This was not for want of trying. After Akasaki's presentation, Nakamura went up to the professor to introduce himself. The old man was chatting to an American colleague. Nakamura waited for a break in the conversation to exchange business cards, as is the Japanese way. When the opportunity arose, Shuji stepped forward, bowing politely, his card thrust forward with both hands, saying, "My name is Nakamura." But to Shuji's surprise, the old man completely ignored him and continued talking to his colleague. Thinking that Akasaki hadn't heard him, Nakamura waited for another pause, then introduced himself again, this time in a louder voice. And for a second time, Akasaki snubbed him.

Such rudeness is hard to excuse but easy to explain. After all, Akasaki was the acknowledged pioneer in the gallium nitride field. It was he who had kept the nitride flame alight for almost twenty years, since Pankove and Maruska had been forced to forsake it. He and his faithful student Amano had made the key breakthroughs, the buffer layer and the positive-type material. Now here he was, an eminent (about to become emeritus) professor at Nagoya University, one of Japan's most prestigious universities, a keynote speaker at a workshop that could have been specifically designed to showcase his life and work. And yet he had been upstaged by a man thirty years his junior, a graduate of a jumped-up technical college somewhere out in the sticks, who worked at a penny-ante company nobody had even heard of, and who had only just started publishing papers. Well, who wouldn't have been offended?

For Nakamura had well and truly stolen Akasaki's show. Shuji had presented papers on his homegrown two-flow MOCVD and his most recent discovery, his much simpler and more effective method for making high-quality p-type GaN. The latter in particular had created quite a stir. Nakamura's contention that getting rid of hydrogen was the solution to the p-type problem provoked one member of his audience to jump up and exclaim, "No, you're wrong!" Somebody else, a researcher who had just published similar results, shot back that Nakamura's theory was in fact correct. Then, to the bemusement of the young Japanese, a ding-dong argument ensued. It was his first experience of the robust nature of American academic discourse.

Shuji was relieved to discover that his English, while not great, was up to the task of presenting his results. To begin with, he read from a prepared script. But this stilted manner of delivery was not to his liking. From the second time on, he spoke off the cuff, a style that suited him better and, he noticed, went over well with his audience. To cap it all, Shuji gave a talk on his blue LED. Given that the device did not emit much light, he was not inclined to regard it as a particularly big deal. He rehearsed the various stages of its development and discussed the theoretical background. Then he mentioned that the LED had a lifetime of more than one thousand hours. No sooner had Nakamura let slip this fact than the audience was on its feet, clapping loudly. Invitations to give talks at other conferences were soon forthcoming. A scientific star had been born.

For years afterward Akasaki would refuse to believe that Nakamura had achieved his breakthroughs working on his own. When I interviewed him in 1995, the old man was still muttering bitterly that Nakamura had behaved in an underhand manner, sending Shiro Sakai as a proxy to his lab at Nagoya to ask Amano all sorts of detailed questions while he, Akasaki, had been away. Around the snobby corridors of Japanese academe, rumors spread that Nakamura was having his wafers grown for him by Sakai, the MOCVD expert at Tokushima University. But, as we have seen, Nakamura had long since broken with Sakai after the latter turned down his proposal to collaborate. If Sakai had pumped Amano for information, if he was secretly growing gallium nitride, then it was for Sakai's own use, not Nakamura's.

In addition to kudos, Nakamura also received another unexpected boost to his morale at the St. Louis workshop. Until then he had been laboring under the misapprehension that the zinc selenide people had beaten him to the punch. Had 3M not already made a bright blue laser? Now he discovered that the lifetime of the ZnSe device was very short. How short? he asked. Less than a second, came the answer. This was common knowledge in the field, but Nakamura had only read about the development in a Japanese newspaper. As usual, the article had waxed lyrical about the positive aspects of the announcement while glossing over the negative ones.

Shuji was reassured: his gallium nitride device might be dim, but since its lifetime was far longer, its potential was much greater. He left St. Louis in high spirits. Nakamura was buoyed by having met others who shared his dream and who had cheered him on. Now, revitalized, he was determined to pursue the dream all the way to its conclusion.

*   *   *

On his return to Anan, Nakamura embarked on the final stage of what he would describe as his climb to the top of Mount Fuji. Thus far, he had managed to fabricate a simple LED. The result, as we have seen, shone with a dim violet-blue light. Now, in order to make a bright, pure-blue LED, he would have to take two further steps.

First, in order to make the light bright, he would have to build a more complex structure, called a double heterojunction. As we saw in chapter 2, this is a fancy-sounding name for a simple concept. Your basic LED consists of just two layers of the same material, differently doped to make one negative and the other positive. This is called a homojunction, the light being emitted at the interface of the two layers. In a double hetero-junction, aka heterostructure, the two layers confine a third, extremely thin layer. This traps the electrons and the holes. Once they fall into this layer, they cannot escape, thus greatly increasing the chances that they will combine with each other to create photons, i.e., light.

Second, in order to make the light pure blue, as opposed to violet-blue, he would have to prepare alloys that incorporated another material, indium, whose slightly narrower bandwidth would produce light of a longer wavelength. The problem was that thus far, no one had been able to make indium gallium nitride of sufficiently high quality for practical use. Following the growth of the world's highest-quality gallium nitride and the production of superior p-type GaN by thermal annealing, growing indium gallium nitride would be Shuji's third major breakthrough.

But even as Shuji girded his loins to meet the challenge, he came under renewed pressure from the company. Nichia's president unexpect-edly insisted that Nakamura should commercialize his dim violet-blue prototype, arguing that, even if it was only 50 percent brighter than sil-icon carbide blue LEDs, they could still sell it. Develop a product, Eiji Ogawa ordered, via written instructions as usual. Stepping up this unwel-come stress, the manager of Nichia's development department and two of his section chiefs began coming one after another to Nakamura's lab to pester him. "You know how much money the company has invested in your work?" they would ask. "Develop a product, and contribute to sales!" For the next month or so they kept harassing him with such demands. "The enemy is not outside the company—he's inside," Shuji used to complain to his juniors during those trying times. He strongly felt that the biggest obstacle was not technological barriers but instructions from the president. Nonetheless Nakamura stood firm. His job was on the line, but he did not care. Paying no heed to the noise from his bosses, he began work on developing indium gallium nitride alloys.

At this time, Takashi Matsuoka, a Japanese researcher who two years earlier had done some preliminary research on growing InGaN crystals at telecom giant NTT, contacted Nakamura. Matsuoka had been excited to learn of Shuji's success in producing high-quality p-type GaN by thermal annealing. He wanted to visit Nakamura to discuss details of this work. Concerned that the NTT man might let the cat out of the bag about what Nichia was doing, Eiji told Shuji to discourage him from visiting. But Matsuoka persisted. He came to Nichia, met Nakamura, and gave a lecture about his work on growing InGaN. However, his material was of poor quality and would not emit light. There was little that Nakamura could learn from his results. Matsuoka returned to Tokyo, where, ironically, he was reassigned to work on zinc selenide. No ignoring your boss's instructions at NTT.

Learning how to grow high-quality InGaN took Nakamura about five months of trial and error, from February to June of 1992. The main difficulty was that the indium gallium nitride layer—which in its ultimate, few-atoms-thick form is known as a quantum well—has to be grown at a much lower temperature than the confining layers of gallium nitride on either side. The bonds between the indium and the nitrogen are very weak. Increase the temperature too quickly and the indium atoms disassociate themselves from their nitrogen neighbors. How to "jump out of the well," as it would later come to be known, and move on to grow the next layer, upping the temperature without destroying the thin layer of InGaN in the process? That was the daunting challenge that gallium nitride researchers faced.

It was at this final hurdle that Akasaki and Amano fell. As we have seen, at the Materials Research Society meeting in Boston in December 1991, the Nagoya University pair had announced a double-heterostructure blue LED. It emitted about ten times more light than the commercial silicon carbide blue LEDs made by Cree, but was still by no means what you would call bright. In attempting to add the crucial element, indium, however, Amano came unstuck. He himself would much later ruefully acknowledge his "great mistake": "I tried the experiments with the thought that it would be difficult, really hard, perhaps impossible. Because it was generally said that gallium nitride and indium nitride would not mix, like water and oil [a phenomenon known in the jargon as *spinodal decomposition*]. . . . But in the nature of things, if you try something suspiciously, you will fail in it."

Nakamura solved the disassociation problem in two ways. First, by brute force, turning the indium tap on his system all the way open, using ten times as much indium as would turn out to be needed, trying to get at least some of the stuff to stick. Second, by guile, adding an extra "blocking" layer to cap the indium gallium nitride layer, thus preventing the material from disassociating. Highlighters poised, students of crystal growth would subsequently scour his papers on quantum well fabrication for mention of blocking layers. But Nakamura would never publicly discuss this delicate piece of high-tech sleight of hand.

In September 1992 Nakamura succeeded in fabricating a double-heterostructure blue LED. Its wavelength was still too short to qualify as true blue. By the end of the year, he had made a few adjustments to the growth program, increasing the amount of indium and reducing the thickness of the active layer. This time, there was no doubting the result.

Much as the power of automobile engines is still rated in terms of the equivalent number of horses, in the lighting industry, lamps are still quaintly measured in terms of their putative candlepower, in units called candelas. Thus far, the output of blue LEDs had been given in millicandelas, or thousandths of a candle. Now, for the first time, Nakamura's device crossed the line into the candela class. It shone with a dazzling sky-blue light, a "furious cerulean" in the eloquent phrase of *Scientific American* editor Glenn Zorpette. It was a hundred times brighter than commercial silicon carbide blue LEDs, bright enough to be clearly seen in broad daylight.

In a word, it was brilliant.

Shuji had gone far enough with his R&D. He had solved all of the previously insurmountable problems. Now, he figured, it should be possible to make a product. And he was confident it would be a product that Nichia could sell. He felt like he was standing on top of Mount Fuji.

In the years to come, there would be many who would try to belittle Shuji's achievement. Akasaki and Amano had made the hard yards, such detractors would argue, saying all Nakamura did was connect the dots. Umesh Mishra, Shuji's future colleague at the University of California at Santa Barbara, begged to differ. "What Akasaki did was pretty cool, but what Shuji did blew the thing wide open—that's the difference."

Herb Kroemer had little time for such quibbles about whose contribution was more important. "There are always people who belittle everything," the Nobel Prize winner chuckled. "I'm sure there were other people involved in the development of the nitrides . . . but Nakamura was the one who made the breakthrough, period. End of statement. And no one else was near that breakthrough. Of course the fact that he was in an industrial environment actually helped in this particular instance, because it did then lead to a product."

Indeed it did. And not just one product, either, but a whole string of them. As Steve DenBaars, another of Nakamura's future colleagues at UC Santa Barbara put it, "When you saw how bright that first blue LED was, at least if you were an LED person, you realized that this was going to open up a lot of new markets."

$$* \quad * \quad *$$

On November 12, 1993, after a year spent frantically gearing up in secret for commercial production, at a press conference in Tokyo, Nichia announced the world's first bright blue LEDs. Lots of journalists showed up for the announcement, ensuring good coverage in the Japanese press. The world's initial reaction was one of outright incredulity. "We got a lot of phone calls, from about fifty companies, but most people didn't believe it," Shuji recalled, laughing loudly. "Because Nichia was a small company with no previous history of involvement with semiconductors. The only people who believed it were the ones who had read my papers. I'd published about fifteen of them by that time."

Gerhard Fasol, a semiconductor researcher then working at the University of Tokyo, was one of the first believers. He wrote an article describing Nakamura's breakthrough for the journal of the German Physical Society. The article was rejected. "The editor explained, I can't print that," Fasol laughed. "The experts in Germany say that it can't be true."

The entire compound semiconductor industry had been taken by surprise. As one Toshiba researcher described it to me, "Everyone was caught with their pants down." Once the disbelief had subsided, however, orders for bright blue LEDs soon started pouring in from all over the world. Early Nichia customers included makers of traffic signals and electronic billboards.

Nakamura continued to push the edge of the envelope. He kept making breakthroughs one after another, pumping out a seemingly never-ending stream of papers and patents. In May 1994 Shuji demonstrated blue and blue-green LEDs capable of emitting two candelas, double the brightness of his original devices. The following year Nichia commercialized bright emerald green–light emitters, the first really green LEDs. In September 1995 Shuji announced the first quantum well–based blue and green LEDs. These featured a brightness of up to ten candelas.

Also in 1995, at Nakamura's suggestion, the company developed white LEDs that worked based on complementary color: sticking a yellow phosphor (Nichia's stock-in-trade) in front of a bright blue LED converted its light to what the human eye perceived as white. Wavelength conversion was an obvious idea: fluorescent lights and black-and-white televisions work on more or less the same principle. But, as we shall see, it opened up huge new markets, including—ultimately—general illumination.

By April 1994 Nichia was manufacturing one million blue LEDs a month and selling them for five dollars a pop (about twice the price of red LEDs of similar brightness). The company had built a six-story factory and hired hundreds of employees to man its new LED production lines. Five years later, the number of employees had doubled, from 640 in 1994 to 1,300 in 1999. By then the company was reportedly producing thirty million blue and green LEDs a month, selling them for $1.50 each. Sales had soared, doubling to $350 million in three years, from 1996 to 1999. Profit margins topped 50 percent.

Meanwhile, nitrides had become a white-hot research topic world-wide, with researchers of all stripes dumping zinc selenide and flocking to jump on the GaN bandwagon. "In industry I see everybody and his brother working on gallium nitride, and at universities almost every colleague I know is working on it," Fernando Ponce told me. A former Xerox PARC researcher, Ponce is now at the University of Arizona. Nakamura used to send him materials for analysis.

"You look at how much effort is going into nitrides now, you pick up *Applied Physics Letters* and a quarter of the papers are nitride papers," Gerald Stringfellow, a University of Utah materials scientist and former Hewlett-Packard researcher told me in 1999. "It's astonishing, absolutely astonishing."

At Agilent (formerly part of Hewlett-Packard)—the preeminent US maker of high-brightness LEDs—R&D manager Roland Haitz urged his boss, "We need a big new program, we need to put a ten-PhD-plus technician team on this thing." Every university in the United States, large or small, seemed to have at least one nitride program. Professors like Ted Moustakas advised their students that a training in nitrides would stand them in good stead throughout their career.

Herbert Kroemer well remembered seeing his first bright blue LED, while on sabbatical in Stuttgart, in the lab of a colleague. "This thing lit up, and I was utterly astonished." Kroemer asked his colleague, "Well, do you think you're going to be working on this? And the latter replied, 'Dr. Kroemer, we have no choice!'"

Kroemer also recalled seeing Nakamura give an invited talk to the German Physical Society in Berlin. "He had a little computer-driven display panel. And I said to the guy who was sitting next to me, What we are seeing here today is the beginning of the end of the lightbulb."

A series of international conferences on the nitrides were held. The first one, in Boston in 1995, was organized by Fernando Ponce. It attracted about 350 attendees. By the third conference, held in 1999 in Montpelier, France, attendance had swelled to over a thousand.

At such conferences, Shuji was the keynote speaker. His sessions were typically standing room only, with crowds overflowing into the hallway. "In the nineties he was the god at every conference we ever went to," crystal grower Warren Weeks remembered. "Everybody would just show up for what we called the Shuji Light Show, because he was always dazzling the crowd with the new products. And he was clearly one, two, three years ahead of everyone in the room, even the best scientists who had flashed blue in their labs. You'd go to the conference, you'd be blinded, and you'd be like, How far ahead is he?"

"Most conferences you go to, you hear a speaker, and it's the same talk he gave at a conference six months ago," recalled Robert Walker, then working for the US MOCVD equipment maker Emcore. "But this was a guy, he'd stand up, and at every conference he had a new break-through: We did 1 milliwatt, we did 5 milliwatts, we did green, we did orange, we have this laser with a ten-hour lifetime, with a thousand-hour lifetime. During that period, for years, he was just incredible."

Shuji's lead was unprecedented. Eminent figures in semiconductor light emitters like Stanford's Jim Harris, who had been active in the field for thirty-odd years, were bemused. "I've never seen anything important that wasn't reproduced with a week or two," Harris told *Scientific American*. "And people are still having trouble reproducing what [Nakamura] did years ago. He's just miles ahead of everybody else."

For Americans, this was galling. "I can't tell you the frustration that has existed in the US scientific community over trailing this small Japanese company," Daniel Dapkus, an engineering professor at University of Southern California, told the *New York Times*. Strenuous efforts were initiated to catch up with Nichia by US, Japanese, and European governments. These typically took the form of research consortiums, with firms joining forces, funded by taxpayer money. The largest US consortium, dubbed Blueband, included Hewlett-Packard, SDL, and Xerox.

For veteran observers of the US semiconductor industry, it was a bizarre spectacle. Throughout the 1980s industry leaders had been complaining about Japanese companies, how they colluded under the leadership of their government to catch up and overtake US rivals. Now, in the 1990s, here were the Americans doing exactly the same thing. They were pooling their efforts, supported by the Pentagon to the tune of $10 million a year, trying to catch up with a lone Japanese researcher. Predictably, resources were misallocated. Blueband was a miserable failure.

Nakamura's biggest coup was developing a blue laser diode. This was something that many people had said would simply not be possible given how riddled with micro-cracks gallium nitride crystal was. For one thing, to amplify light, a laser needs a more complicated structure than an LED; for another, you have to pump a laser with far more current. The structural defects in the material should have scattered the light all over the place, preventing optical amplification. When subjected to high current, the defect-ridden layers should have caused catastrophic failure within a matter of moments.

It seems strange now, but back in the mid-1990s, making a blue laser was seen as being more of a big deal than making a blue LED. The reason is that, whereas it was not easy back then for nonspecialists to imagine all the applications that would emerge for bright blue LEDs, it was very clear indeed what the big application would be for a blue laser: data storage.

Infrared lasers were already in widespread use for CD players and drives. Red and orange devices were on the way for DVD players. With its much shorter wavelength, blue would increase the amount of data that could be stored on and read back from a disc by more than six times. That is, from 4.7 gigabytes for today's DVDs to about 25 gigabytes, enough for about nine hours of video, that is, three high-definition movies or video games. That was why every consumer electronics company—notably Sony—and data storage specialist—notably IBM, at its Almaden laboratory in San Jose—were pouring money and resources into developing a blue laser. Worldwide, maybe twenty of the top companies in the industry were working on the problem.

Shuji beat them all to it. In 1996 he unveiled a prototype gallium nitride violet-blue laser at an international semiconductor physics conference in Berlin. Ever the cheeky showman, during his presentation he used the laser as a pointer. In Berlin he was invited to a dinner in his honor. The guests included Klaus von Klitzing, who, unbeknownst to Nakamura, was the winner of the 1985 Nobel Prize for Physics. Toward the end of the meal, Shuji, who had been chatting away with von Klitzing, asked him what he did. Gerhard Fasol, who had organized the dinner, told Nakamura that he was talking to Germany's only winner of the Nobel Prize for Physics. "WHAT?!" Nakamura blurted, "You won the Nobel Prize?—What did you win it for?" And Fasol had to laugh, partly at Nakamura's naivete, but mostly because asking such an eminent scientist to justify the reason for his eminence is just not done, certainly not in Germany. "Most people von Klitzing can just brush away, telling them to go and read his papers. But Shuji Nakamura is one of the few people in the world that he cannot brush away like that." For Nakamura was, after all, the inventor of the bright blue light emitting diode and the blue laser. And, as such, not a man to be taken lightly, even by a Nobel Prize winner.

The lifetime of this first blue laser was just thirty-five hours, far too short for a commercial product. To improve the longevity would require a major reduction in the number of defects in the crystal. By 1996 Nakamura was traveling a good deal, attending conferences and workshops. He could no longer spend all his time in the lab. He was now the leader of a team of young researchers at Nichia.

Being out in the world had its advantages. At a panel session he was chairing at a Japanese conference in spring of 1997, Shuji heard a researcher from NEC describe how his group had used a technology to reduce the defect density by a thousand times. Known in the jargon as lateral epitaxial overgrowth, or LEO, it first been used to improve the quality of gallium arsenide solar cells grown on silicon wafers. Shuji's friend Steve DenBaars of the University of California at Santa Barbara who was sitting next to him remembered what happened next.

"At the end of the session, Shuji jumped up, ran to the other end of the room, and got on the phone. When he got off the phone, I asked him why he was so excited? He said, 'We must begin work immediately on LEO!' That's an example of him hearing an idea and being able to implement it into a product. Because the NEC guy was not talking about putting LEO into a product, he was just doing it for the sake of the science."

By the end of the year, after several months of frantic work, Shuji and his group at Nichia had applied LEO technology to make an improved blue laser that operated for one thousand hours. Nakamura announced the result at an international conference on nitride semiconductors held in Tokushima, with many of his friends in attendance. He would never forget how proud he felt on that occasion. Rightly so—the world had come to his doorstep to hear him speak.

The way to commercialization was now clear. By June 1999, when I visited Nichia for the third and final time, the company had recently begun sending out sample shipments of blue-violet laser diodes for use in high-definition DVD players. At the time of writing, more than seven years later, Nichia is still the only company shipping blue lasers in commercial quantities.

✳ ✳ ✳

On arriving at Nichia in 1999, I marveled at the changes that had taken place since my first visit to the company, five years earlier. In particular, the massive new LED factory that had recently shot up, dwarfing Nichia's headquarters building. Here was positive proof of the scale of Nakamura's achievement—and the company's new-found prosperity. Nobuo Ogawa's gamble on Nakamura had paid off, big time.

Summoned by the girls at reception, Shuji came bounding out to greet me in his company overalls, grinning hugely as always. We chatted about his achievements. He seemed to feel that the era of discovery was almost over. The big breakthroughs had all been made. Shuji's main purpose had been to develop bright blue LEDs, and that had been achieved. Blue, green, and white LEDs had been commercialized, and blue-violet lasers were at the sample shipment stage. What was left was mostly mopping up operations—improving the material quality still further, increasing the efficiency of light output. Lots of product development and applications work; more engineering than science.

On both of my previous visits, I had interviewed Shuji by himself, one-on-one. On this occasion, though, Shuji was shadowed by another employee, an older man whose business card identified him as a marketing executive in Nichia's sales promotion department. In my experience, marketing people tend to be voluble: it can be hard to shut them up. With this character, however, it was the opposite: I couldn't get him to say anything. I would ask some innocuous question about how Nichia's LED business was doing, and he would just snap, "No comment." Nor would he allow Shuji to answer.

The situation was almost farcical, though at the time I did not find it particularly amusing. I had after all journeyed many thousands of miles to interview Nakamura for a corporate profile in the global edition of *Forbes*, one of the world's most widely read business magazines. The article—a cover story as it turned out—would undoubtedly provide favorable publicity for Nichia. But here was this minder who was refusing to provide answers to even the simplest inquiries. All of this, I could see, was affecting Shuji. He was literally squirming with embarrassment.

What I discovered later, long after Shuji had left Japan, was that Nichia had become concerned, knowing that Nakamura was by nature a very open sort of person, that he might reveal company secrets. So the tight-lipped company had done its best to keep Shuji quiet, assigning him a corporate watchdog to make sure he kept his mouth shut. This was offensive to Shuji and, at least as far as I was concerned, totally unnecessary. There was little likelihood that I would ask the kind of detailed question that might in any way jeopardize corporate secrets. For his part, as a loyal company employee, Shuji was well aware of what he should and should not say. There was no chance of him blurting out something confidential.

In retrospect, the only scoop I could have come away with was the fact that Shuji was thinking of leaving Nichia. But at the time, he was still several months away from making up his mind about this. When I saw him, unbeknownst to me, he was contemplating taking up a job offer from the US company Cree. A few weeks later, he would travel to the University of California at Santa Barbara for the first time. There, his friend Steve DenBaars would persuade him that joining a rival company would not be a good idea. Better to become a professor: less chance of being sued that way. Little did they know!

That October, four months after my visit, Shuji was still insisting to my friend Dennis Normile, in an interview for *Popular Science*, that he had no intention of leaving Shikoku. "The calls [job offers] come in, but I turn them all down," he told Dennis with a shrug. "This is like family here; I couldn't say good-bye." Dennis concluded his profile with Shuji's assertion of loyalty. Unfortunately for Dennis, it was not true. By the time the story appeared in the magazine the following February, Nakamura was long gone.

Shuji quit Nichia in December 1999 to join the faculty of UC Santa Barbara as a professor in the materials department. My editor at *Forbes* in New York was understandably keen to know why he had gone. In January I e-mailed Shuji saying that I expected his departure from the company, where he had worked for almost exactly twenty years, was because he wanted to pursue the scientific aspects of his work in an academic context. Also, that being a very open sort of person, it was difficult for him to have good scientific interactions while at the same time maintaining commercial confidentiality.

Ever the good correspondent, Shuji replied immediately: "Your guess of the reason why I leave Nichia is right." But in fact, I was only partially right. The real story of what had actually happened to Nakamura at Nichia, of how shamefully his company and, in particular, its president had treated him, would not emerge for some time.

How this came about, how Nichia managed to shoo away the goose that had laid the golden eggs, is the subject of part 3. Before dealing with his departure for the United States and its extraordinary legal consequences, we must bid Shuji adieu for now and turn our attention to some of the initial ways that the rest of the world has found to apply his brilliant devices.

But before doing that, however, we must first look at the history of blue LED development at Nichia's American archrival, Cree, and outline the origins of the relationship between Nakamura and the US firm, which would be the genesis of his legal battles.

CHAPTER 6

John Edmond, one of the six founders of Durham, North Carolina, based Cree Research, recalled that it was on December 12, 1993, when he and his colleagues saw the article in *Electronic Engineering Times* reporting the sensational news that Nichia had achieved a candela-class bright blue LED. "We were like, Holy shit!" Edmond said, sarcastically, "this is a great Christmas present!"

As by far the world's leading maker of blue LEDs, Cree was the company most affected by the Nichia announcement. In times of crisis, you have to move fast. Two days later, Edmond and his colleagues were sitting in a conference room in a Tokyo hotel. Facing them across the table were Nichia's top management and Shuji Nakamura. Edmond had managed to organize an audience with Nichia on short notice because he knew Nakamura from going to the same conferences together. The Americans were eager to see Shuji's new device. When one was produced and illuminated, the visitors gasped in astonishment. They were hardly able believe their eyes. "It was fifty times brighter than our brightest stuff."

Edmond and company countered by proposing a collaboration with their Japanese counterparts. Cree was also working on gallium nitride. Within six months they would have a production-grade device that would be just as bright. It made sense to work together. The Americans could supply Nichia with silicon carbide substrates. But the Nichia folks were simply not interested. "They were like, *gaijin* [foreigner], go home. It was just, How many LEDs do you want to buy?" None, the Americans spluttered, we don't want to buy any. "But they didn't care what we said, they didn't care that we were working on nitrides."

The Cree team returned home stunned by what they had seen and heard. At the same time, however, the flat refusal on the part of the Japanese to collaborate also served to pump the Americans up, to stiffen their resolve. "So it was a shock on the one hand, but it was like, we're going in the right direction, we're going to crush them in the end, we've just got to keep working," Edmond recalled. "We were young and stupid . . . and very confident." A resilient ego was ever a necessary precondition for successful entrepreneurs.

Edmond had predicted that Cree would come up with a competitive LED in six months. In fact, it took the Americans a year and six months to release their first gallium nitride product. "That was in June '95, and it was a crappy device. It was maybe half the brightness of the Nichia one, and Nichia had made improvements since then. But at least half the brightness was better than one hundredth or whatever it was."

Meanwhile, in their determination to catch up with and crush Nichia, the Americans had also devised another strategy. They would try to poach Nichia's star researcher. At the end of 1994, Edmond sent his Japanese counterpart a letter inviting him to come and join them. In it, he wrote that Cree would double Shuji's salary. More importantly, the US company, which had gone public the previous year, also offered him generous stock options. Had Nakamura accepted the options then, they would in due course have made him a very rich man.

Trouble was, Shuji did not know what a stock option was. This is not so surprising, given that stock options would be illegal in Japan until 1997. He went home and asked his wife whether she had any idea what "stock option" meant, but Hiroko didn't know either. At the time Nakamura was so busy working on developing a blue laser that he neglected to look up the meaning of the term. When Edmond called from the United States to see whether Nakamura was interested in joining Cree, Shuji politely declined his offer.

Over the next six years, Cree would make repeated attempts to persuade Nakamura to leave Nichia and join them. Eventually, at least to some extent, they would succeed.

* * *

Cree Research was founded by two brothers, Eric and Neal Hunter. They hailed from the small town of Boone, which is named after the famous frontiersman Daniel Boone. The town is located in the beautiful Blue Ridge Mountains of northwestern North Carolina, a range that includes the highest peaks east of the Rockies. The brothers attended Watauga High School, where they both competed on the same swim team.

On first hearing the name Cree, I imagined that it derived from the Native American tribe of the same name. Many years later I discovered that the Cree live in Alberta, Canada, which is a long way from North Carolina. In fact, Cree is an abbreviated form of the Scottish family name McCree, also spelled McCrae. It happened to be the middle name of the Hunter brothers' late father, a doctor in whose memory they named their venture.

The Hunters both attended North Carolina State University, one of the three schools that form the triangle that gives Raleigh-Durham's Research Triangle Park its name. Neal was a mechanical engineer who, on graduation, got a job as a salesman with a big control systems company. Described by a high school friend as "very smart, very athletic, and very competitive," Neal quickly discovered that he did not like working for other people: at heart he was an entrepreneur.

In his older brother, Eric, the entrepreneurial tendency was even more pronounced. Bob Davis, his professor at NCSU's graduate school, observed that Eric had wanted to form a company from the day he arrived. "You could tell he was an entrepreneur," John Edmond recalled. "He was always on the phone in the lab, talking to brokers, buying and selling stock. There'd be a crystal growth run going on over here, and I'm like, What the hell is he doing—doesn't he have better things to do?"

The specialty of the Davis lab was silicon carbide, a compound semiconductor material. Silicon carbide does not exist in nature; it was first synthesized in 1808. For most of its history this substance, which is made by reacting sand and coke in a furnace, had been used in abrasives. As we have seen, it was the first material in which electroluminescence was observed. Then, in the 1950s, it was realized that in the form of electronic devices like transistors, silicon carbide might be able to go places where plain old silicon could not, withstanding higher temperatures, or operating at higher powers and frequencies.

But how to grow bulk crystals of the recalcitrant material, which you could slice up into wafers and fabricate devices with? This presented a formidable challenge. For one thing, you have to heat the stuff up to searing temperatures. For another, when you heat silicon carbide, unlike most materials, it doesn't melt. It *sublimes*. In other words, the compound jumps straight from being a solid to a gas, just like dry ice. Then, when it crystallizes, silicon carbide has 177 different structures that it wants to form. Controlling the growth process so that the crazy compound crystallizes into the structure you want is extremely difficult.

Nonetheless, that is what over ten years Bob Davis and his students at NCSU, notably Calvin Carter, managed to do. They became the first group to grow bulk silicon carbide, ahead of several large corporations, including Westinghouse. The crystal material they made was sliced into tiny, fingernail-sized wafers on which devices could be grown. "We would take the wafers and divvy them up between us," Warren Weeks, who worked at the Davis lab in the early 1990s, recalled. "They were as precious as you could get, we'd break them up into little bitty pieces and try to make them last as long as we could."

In July 1987, after a meeting at a biscuits restaurant in Boone, Eric and Neal Hunter decided to form Cree Research to commercialize the work. They licensed ten patents from the university. In return Cree issued the university shares in the company. The Hunters took out second mortgages on their homes and raised $25,000. As their corporate headquarters, they moved two desks into a poky eight-by-thirteen-foot office.

John Edmond, a member of the Davis group whose specialty was ion implantation—a high-tech method for adding dopants to semiconductors—was the first full-time employee the Hunters hired. "They said, We're going to give you an advance," Edmond recalled. "I said, Man, that's good! Because being a graduate student, I didn't have any money. So they went and maxed out their credit cards and gave me a check for 4.2 months of salary. And that's why I came on board." Two other members of the Davis lab, Calvin Carter and John Palmour, joined shortly afterward.

The rangy, intense Edmond was the only one of the group who was not local. Indeed, so far from being a Tar Heel, he was actually a Northerner, from Cohocton, a town of nine hundred residents in western New York State. A self-confessed propeller head and science nut, Edmond went to the New York State College of Ceramics at Alfred University. Neal Hunter would tease him by referring to his alma mater as "the fighting cement mixers of Alfred." As the winner of a scholarship from the Office of Naval Research, Edmond could have gone to any graduate school in the country, but, attracted by Davis's reputation, he ended up at NCSU. "I was interested in a ceramic material that could be used in electronics. They were working on silicon carbide. I thought, Silicon carbide? That's a ceramic, they use it for abrasives, grinding wheels, sandpapers; if it's a semiconductor, that's perfect, that's what I want to work on."

In 1987 Edmond was twenty-six years old, Neal Hunter just twenty-five. Chances of success in founding a start-up straight out of university are not good. But, as Hunter later put it, "I was too young and too stupid to know." As with every start-up, securing funding proved a major initial obstacle. Cree Research began with four contracts from the Defense Department's Small Business Innovative Research program. But just weeks after Cree's formation came Black Monday, October 19, 1987, the largest one-day crash in the US stock market's history. Venture capital virtually dried up. But the company was rescued from oblivion by an unlikely savior. The Hunters tapped a local McDonald's franchisee who was dating a cousin of theirs. The franchisee brought in some of his friends from McDonald's as angel investors. "They were a great group of guys; they believed in us," Edmond said. "They put the money there. They made gazillions of dollars on Cree, because they put a lot into the company early on."

With cash in hand, the question for the fledgling firm now was, What to make using their silicon carbide expertise? Fancy devices for use in high-power switching and high-frequency communications would take time to develop. In the short term Cree needed a cash cow, a product for which there was an existing market. The Hunters approached Edmond. "They said, OK John, we want you to make a blue LED, that's going to be our first product. I said, What the hell's a blue LED?" Eric Hunter told him not to worry. "It's easy," he reassured Edmond. "It's like making donuts!"

*   *   *

Unlike Nakamura, Cree had chosen to do blue not because silicon carbide was the best material for the application, but because silicon carbide was what they knew how to make. At that time, the reputation of blue light emitting diodes among LED makers was, to put it mildly, not good. "I always considered blue LEDs a black hole," former Hewlett-Packard R&D manager Roland Haitz told me. "Around the late eighties I could be quoted as having said that I was proud that HP had not spent a single dollar on blue LED development." As far as Haitz was concerned, for a light emitter, blue was an "unnatural act."

With word about recent Japanese developments in gallium nitride beginning to filter through, however, that negative attitude was changing, albeit slowly. Edmond recalled visiting George Craford at HP around 1989. "They had one guy who had been working a little bit with blue LEDs, but they were like, These things will never work, you know. And obviously they knew that silicon carbide wouldn't work."

Edmond began his attempt to figure out how to go about making blue silicon carbide LEDs by combing through the literature, to find out who had done the best work. The answer turned out to be the giant German firm, Siemens. To make their devices, Siemens employed a primitive process. From the blowholes of the giant furnaces they used to make silicon carbide abrasives, they would go in and chisel out tiny crystals of dark black material. These crystals would be the basis for making LEDs using the traditional method of vapor phase epitaxy. The LEDs thus produced shone blue, but they were so dim you could barely see them. The devices were sold individually, for scientific purposes, as a spectral reference for blue light. Produced in minute quantities, they were very expensive. Edmond remembered ordering two Siemens LEDs from Hallmark Electronics for seventy-five dollars apiece. In 1988 the Japanese consumer electronics firm Sanyo came out with a silicon carbide LED that was brighter than the Siemens device. It sold for about ten dollars. "We looked at what they were doing, and it was a vapor phase epitaxy process also," Edmond recalled. "We said, This is just stupid, we can't do that—let's do something different."

Cree's experience was with the new high-precision growth method MOCVD, the very same technique that Shuji Nakamura was coincidently around this time learning nearby at the University of Florida. One of the company's first employees, H. S. Kong, a native of Shanghai, had mastered MOCVD while working in the Davis lab at NCSU. In fact, since the precursor materials he was using (silane and methane) do not count as metal organics, the method was really just a variant of vapor phase epitaxy. It required extremely high temperatures of around 1,700 degrees Celsius, higher than the melting point of quartz. Somehow Cree's crystal growers figured out a way to handle this insane process.

The Cree team mustered in-house all the other skills required to succeed in the LED business. Tommy Coleman, one of the company's six founders, was a hands-on equipment builder. Carter was the bulk silicon carbide crystal growth expert, Edmond and Palmour the device fabricators, Neal Hunter the salesman. Starting from scratch, it was a case of believing in themselves, believing that they could make blue silicon carbide LEDs, and then doing what they knew how to do. But it was a bumpy ride. "We had several months, and many times when we couldn't grow a crystal," Edmond recalled. "Then we'd say, OK Calvin—you're not in charge of crystal growth anymore, Neal's in charge! So Neal would go in and he would do just this wild stuff. I mean, we've got pictures of crystals he grew that were great looking; they were dog-shit wafers, but they looked cool. For example, the old egg crystal, it looked like an egg, like it had a yolk. It was unbelievable."

"We would challenge each other. I worked in the polishing and cutting and grinding of the crystals, all sorts of things . . . we did everything. We would interact, we had a lot of passion, which I think is the key to success: we loved what we did, we believed in what we could do, and you'd just go in there and do it. In crystal growth there's a lot of black magic—I mean, just weird things going on."

There would be dark days when, in desperation, they would be forced to fall back on their last resort—the magic power of a rubber chicken. When all else failed, they would grab the dummy bird by its neck and swing it in an incantatory gesture over the machine that they were having trouble with, all the while chanting the mantra, "C'mon big money!" The demons having thus been cast out, the thing would start working again. Or at any rate, the cause of the problem would have been identified. Such fun rituals became part of the close-knit Cree culture.

The team worked together and played together. Out back of their building was a basketball court where they would come off work and shoot hoops or play pick-up games. Here, too, their fiercely competitive nature would be in evidence. "Those guys really want to win," said Warren Weeks, the crystal grower from the Davis group, who worked for Cree in the mid-1990s. They drove themselves hard. Their employees, too: "I was only at Cree for two years," Weeks recalled, "but they were really dog years. It felt like fourteen!"

Working close together under great pressure saw lasting bonds of friendship and respect develop between the founders. "We were like brothers, a really good family," Edmond recalled. "You'd get here early in the morning, you'd eat lunch together, you'd have dinner together, you'd come back here and work till eight or nine o'clock, then you'd go home, and you'd do it seven days a week. There was no day off, for years." Like Nakamura, they were driven. "You've got to be self-motivated; you've got to love what you do."

This relationship would stand the test of time. When I met Edmond at Cree in late 2005, he was just about to go on vacation with some of the other founding members. He had recently bought land to build a house out at Colvard Farms, a rustic six-hundred-acre parcel of land nearby that Neal Hunter had converted into an upscale subdivision, characteristically naming the development after his maternal grandfather. "Neal's out there, Calvin's out there, I'm out there. We really are a close group. We were in school together, we grew the company together, we're all going to be living next to one another—it's going to be kind of fun."

*   *   *

In October 1989 Cree introduced its first commercial blue LED. "It was much better than the Siemens and Sanyo devices, and we were selling it for a dollar," said Edmond. By this time, however, the company was almost broke, with just a month's worth of cash left. They were "about ready to starve," as Edmond put it, when an unexpected savior arrived on their doorstep in the shape of representatives from the giant Osaka-based trading company Sumitomo Corporation.

Japan's *sogo shosha*—general trading companies—comb the world looking for unique new products to introduce to the Japanese market via their extensive distribution networks back home. Blue LEDs were not exactly unique, but as we have seen, there were only two rival products, both of which were much more expensive than Cree's dollar-a-pop devices and not nearly as bright. In 1990 Sumitomo signed up the tiny US firm—Cree then had only about twenty employees—to a Japanese distributorship agreement worth one million dollars. It was a match made in heaven, a relationship that flourishes to this day. Sumitomo would become Cree's biggest customer, accounting for about a third of the company's overall sales. That initial agreement would be followed by a series of others, the most recent at the time of writing being a 2005 purchase of a massive $200 million worth of bright blue LED chips.

Some huge ironies here. As it happened, in 1990, the year that Cree and Sumitomo got together, I visited Research Triangle Park. I was there researching an article on what former Harvard professor Robert Reich had called "the rise of techno-nationalism." That is, a desire to exclude Japan as punishment for what critics saw as Japanese companies having taken a free ride on US science and technology. The Japanese were manufacturing high-tech products, such as computer memory chips, undercutting American firms and driving them out of business. US R&D consortia based on the supposed Japanese model of government-industry collaboration were springing up. One was Sematech, a chip-manufacturing equipment consortium based outside Austin, Texas. This nationalistic organization would not even entertain visits by Japanese journalists, let alone membership by Japanese companies. Predictably, however, techno-nationalism was most prevalent in Washington, DC.

What I discovered at Research Triangle Park, only a hundred or so miles south of the nation's capital, was a very different story. There, in a state that then ranked thirty-fourth in terms of per capita income, the Japanese were welcomed with open arms as a source of much-needed investment and jobs. Several Japanese firms had set up shop in the park. They included Mitsubishi Semiconductor, a chipmaker, and Sumitomo Electric, another branch of the conglomerate, which produced optic fiber. I also found other forms of benign Japanese interest. They included the sponsorship by Kobe Steel of a chair at North Carolina State University. The chair was then occupied by a professor named . . . Bob Davis.

In those days, there was not yet much evidence at Research Triangle Park of an entrepreneurial culture. Cree would be North Carolina's first great start-up success. And it was Sumitomo, a Japanese firm, which initially underwrote that success.

A final irony, for those disposed to notice such things, was the fact that in the United States, the land of the rugged individual, Cree was as we have seen very much a team outfit, with members dividing up various tasks among themselves. Whereas Japan, a notoriously group-oriented country, had produced Shuji Nakamura, a lone researcher working independently, doing almost everything by himself. So much for national stereotypes.

*   *   *

The initial customers for Cree's blue LEDs in Japan were mostly themselves LED makers. For example, Stanley Electric built the first large full-color LED display screen, which it unveiled at the 1992 Electronics Show in Tokyo. The screen was not very bright, but it was good enough for indoor applications. This display was in some sense an early vindication for Jim Tietjen's vision at RCA back in the late 1960s of an LED television that could be hung on a wall. A few such screens were sold, one of them to the new international airport in Seoul.

Pachinko parlors, those gilded palaces of (venal) sin that are to pinball-loving Japanese what slot-machine casinos are to Westerners, would become one of the largest markets for Cree's chips as the signage on the parlors transitioned from gaudy neon to even brighter colored LEDs. But that would come later.

The reason Cree was able to crack the Japanese market in the first instance was that, Sanyo having dropped out, there was simply no other source of blue LEDs. The American devices were it. In April 1992, however, Edmond received his first intimation that this happy monopoly was not indefinitely sustainable. He was one of the handful of attendees at the first gallium nitride workshop in St. Louis. This, as we have seen, served as Nakamura's US coming-out party. "Shuji showed his LEDs and he compared them to ours, because there wasn't anything better. His were still way down here [in terms of brightness], and ours were way up there, and it was so funny because our LEDs were the benchmark, an indirect-bandgap semiconductor."

It was not so funny the following year, when Edmond and Nakamura met up again, at another conference. "He said, I've got something that I can't talk about because my boss won't let me, but it'll blow your stuff away. And I go, Yeah, sure Shuji. And he goes, I'm telling you."

Edmond took the hint. Especially after he bought some Akasaki-style gallium nitride MIS devices made by Toyoda Gosei that were several times brighter than Cree's best blue LEDs. A grant from the Defense Department to work on nitrides for blue lasers helped kick-start gallium nitride research at Cree. "We were working hot and heavy on gallium nitride, because it's a direct-bandgap material, so it's got to be better." Then, at the end of 1993, came Nakamura's shattering announcement of candela-class bright blue LEDs.

Luckily for Cree, there would be a period of grace in which to sell its dim-but-cheap blue silicon carbide LEDs while Nichia ramped up production volumes. During this time, at five dollars each, bright gallium nitride devices were still comparatively much more expensive. In 1994 the US firm shipped seven million silicon carbide devices, accounting for 99.9 percent of the market. Applications ranged from Stanley's indoor displays to the indicator lights on Crown audio amplifiers.

By June 1995 Cree had developed its first GaN devices. In the rush to meet the challenge from Nichia, it was suggested they abandon their trademark silicon carbide in favor of the sapphire substrates that everyone else was using. Edmond was determined to stay with SiC. It was partly a matter of pride, partly the not-invented-here syndrome. He would be damned if Cree's devices were just like Nichia's. Grafting the two materials onto each other turned out to be a nightmare—though the lattice match between gallium nitride and silicon carbide is much closer than between gallium nitride and sapphire, GaN will not grow on SiC substrates—but eventually Cree's researchers discovered a wonderful trick that saved the day. For years afterward Cree would continue to describe its products as "silicon carbide LEDs." In fact, however, they were really gallium nitride devices grown on silicon carbide substrates. Cree was the first, and would remain the largest, US manufacturer of gallium nitride devices.

Cree's first-generation gallium nitride devices had an unfortunate tendency to burn out. By 1996 Edmond had fixed the problem. In that year the company got a really big break. The president of the German carmaker Volkswagen decided that, in order to differentiate its vehicles, the dashboard lighting on them should be blue. He approached Siemens, who in turn tapped Cree, who the Germans knew through the silicon carbide connection.

In addition to differentiation, LEDs offer other advantages for car interiors. They let carmakers shrink the thickness of the instrument cluster in the dashboard. Incandescent bulbs are intrinsically big; they also emit a lot of heat. Thus, if positioned too close to the plastic parts they are illuminating, over time the plastic distorts and melts. Some space must be left between bulbs and instruments. But as carmakers try to cram in ever more features, airbags and what have you, space becomes a premium. Saving an inch or so in the thickness of the dashboard is thus a significant benefit.

Interior lighting for cars was a huge application; in fact, it drove the market. By 1997 Cree was shipping 75 percent of its LED production to Siemens for use in VWs. The devices, which Cree still makes today, were simple gallium nitride blue light emitters, blue being VW's signature color. Cree focused on producing these low-brightness devices because that was what their customer wanted, and the US firm was keen to comply with VW's requirement. Meanwhile, in Japan, Nichia had begun manufacturing complex high-brightness blue devices with indium gallium nitride quantum wells. Edmond tried desperately to catch up with Cree's Japanese rival, but to no avail.

"They were always a step ahead. We'd come up with an indium gallium nitride device that was about the same. Then they'd come up with one that was brighter again. Back in the late nineties I gave several presentations at LED conferences, and I always followed Shuji. It really pissed me off; it became a running joke because [the organizers] would always put me right after Shuji. So he would come up with these new devices that would, like, blind you. And I'm sitting back there saying, Oh God, I've got to follow him. And he did it to me over and over again."

During these years, Edmond had many interactions with Nakamura. He made several attempts to hire Nichia's star researcher. But Shuji would refuse, always having an excuse. For example, his wife's mother was elderly, and they couldn't leave her behind in Japan. Then, in October 1999, Nakamura happened to attend a conference at Research Triangle Park. On a visit to Cree, Shuji made it clear that he was unhappy at Nichia. "We were like, Hmm, that could be interesting," Edmond recalled. "Then we started talking seriously about getting him out of there."

✳   ✳   ✳

How things panned out in Shuji's departure from Japan and the subsequent, dramatic developments is the subject of part 3. Then, in part 4, we investigate how the transition from conventional to solid-state lighting will take place, how LEDs will replace the lightbulb.

This is a process in which Edmond is determined to participate. "Ten years from now I want to see the incandescent gone. It's going to have to be cheap . . . but we can get there, it can be done." Now multimillionaires, he and his colleagues no longer have to work, but nonetheless they still do. "Why? Because we love it. I'm more excited about Cree than I ever have been. I want to replace these," Edmond told me, pointing at the fluorescent strip lights in the ceiling of the conference room at Cree's head-quarters in Durham.

For that to happen, considerable increases are required in the conversion efficiency of LEDs, up from around 50 percent today. "I believe 75 percent efficiency will happen in the next five years. We've already demonstrated 100 lumens per watt on a smaller chip." Lumens per watt being the lighting industry's equivalent of miles per gallon.* Today, the vast majority of Cree's revenue comes from such small chips used for applications like cell phone backlights. But the company's goal is to move toward larger "power" chips for use in general illumination. At 70 lumens per watt, LEDs can compete directly with compact fluorescent lamps. At 100 lumens, they can take on fluorescent tubes. "My objective is to get to 100 lumens per watt on a power chip. I think that would set things on fire, personally. That's where it has to be to really *git things crankin'*."

"It's going to be amazing, what's going to happen next—with head-lights in cars, backlights in LCDs, general illumination, I mean, it's just the beginning. So I want to see this continue, our baby [Cree] is only eighteen years old as far as I'm concerned. I've got several years left in me. I'm forty-four, not ready to retire yet. I want to see us at a point where you can go into Lowe's and buy a light that's all LEDs, and it's putting out 1,500 lumens and it's using 15 watts of energy. When that happens, I probably will take some time off."

In 2004 Cree announced that it had officially entered the race to replace the lightbulb. That year, the company posted annual sales of $390 million, with LEDs accounting for 82 percent of the total. The company was worth $1.7 billion and employed over 1,300 people. But it was still true to its roots. "Even though they have 1,400 people over there, they still operate like a start-up," said Bernd Keller, manager of the company's Santa Barbara Technology Center. "They're very opportunistic, very aggressive, very high speed; that has sustained the company and made it what it is today."

In April 2005 Neal Hunter, at forty-three, resigned as president of Cree, "to pursue other opportunities." These, as we have seen, included

---

* A lumen equals the amount of light given out by a one-candela source radiating equally in all directions.

property development. But building and selling houses was never going to be enough to satisfy Neal Hunter's entrepreneurial urges. Six months later, in October 2005, he and three former Cree cohorts announced the launch of a new company, named LED Lighting Fixtures. The start-up's immodest goal: to take on what it saw as the big names in the lighting industry, GE, Philips, and Osram, for a chunk of the $40 billion global lighting market.

"I think it's time to challenge them," said Neal Hunter, who is not known for being the most patient of people. Such firms had no incentive to innovate so long as consumers keep buying lightbulbs. The consumer market for LED lighting was not moving fast enough for his liking. Hunter's new company would help accelerate the transition from light-bulb to light emitting diode. Needless to say, its lighting fixtures would use LEDs supplied by Cree.

In April 2006 the start-up announced its first lighting fixture, a recessed can luminaire that produced, according to independent tests, 73 lumens per watt. The fixture operated using less than 15 percent of the power of a comparable incandescent bulb and 50 percent of a compact fluorescent lamp. Drawing fewer than 10 watts, the lamp was cool to the touch. The company claimed that it could provide fifty thousand hours of light—enough for twenty years under normal usage of five to six hours per day—compared to two thousand by incandescents. By replacing incandescents with LEDs, consumers would save seven dollars per fix-ture in annual energy costs. The fixtures would pay for themselves in less than two years.

"LED Lighting Fixtures' technological advances shatter conventional thinking regarding the projected timeline and quality of light offered by LEDs for general lighting applications," the announcement claimed. "This revolutionary reinvention of light could have global implications for decades to come." Production and distribution of the new fixture, which would be manufactured in China, was slated to begin by the end of 2006.

Part Two
THE FLOODGATES OPEN

CHAPTER 7

The first applications for Shuji Nakamura's bright blue and green light emitting diodes were obvious ones, for the most part extrapolations of existing product trends. Long-lasting, energy-efficient, maintenance-free light emitting diodes had already begun to replace the filtered incandescent bulbs in the red stoplights of traffic signals. Now, LEDs could replace the green go lights, too, cutting the monthly running cost of a set of signals from more than twenty to just a few dollars.* As we have seen, Cree's dim blue LEDs had already been used in large flat-panel displays for indoor use. Now, with the advent of Nichia's new products in 1996, large-screen displays could be made bright enough for outdoor use, too. For example, the seven-story NASDAQ display in New York's Times Square. This monster screen consists of nearly nineteen million LEDs and covers almost a quarter of an acre, "a new icon for the era of the marketplace, a blue-chip supersign on a once gritty corner."

But the arrival in the market of high-brightness blue and green LEDs to complete the trinity of primary colors also had less obvious implications, too. These had to do not with the brightness of the lights but with

---

* In October 2005 Dialight, a leading maker of traffic signals announced that, over seven years, it had used seventeen million LEDs without a single failure in the field.

their unprecedented controllability. Light emitting diodes were semiconductor devices, a nice match for other kinds of semiconductor devices, like microprocessors. LED lighting fixtures and the technology to control them via computers would provide a whole new toolkit of programmable lighting effects with which artists, architects, and lighting designers could create. Among the first to recognize these non-obvious implications of solid-state lighting were George Mueller and Ihor Lys. Over the next decade Color Kinetics, the company these two twenty-somethings founded in 1997 to pursue their vision, would blaze a trail for the revolution in semiconductor illumination. Initially this would be colored lighting, for decorative purposes. White would come later.

As you might expect with a disruptive technology, Mueller and Lys did not come from a lighting industry background. In fact, their previous experience was in the robotics laboratory at Carnegie Mellon University, where both had been graduate students in electrical engineering. There, they had been involved in the development of autonomous robotic vehicles, driverless Humvees that, at certain times of year, would careen off into the Nevada desert vying with their robotic rivals for a $2 million prize from the Pentagon. Riding shotgun in one of these autonomous vehicles could be terrifying, as Lys vividly recalled. "Computers do not have a fear term that gets put into their speed calculation," he said. "If the computer knows that it's right, it's going to floor the accelerator and go as fast as it wants. And it knows that the vehicle won't tip over, so it knows exactly how far it can turn the wheel."

In this work, LEDs were one of those technologies that just kept popping up, in the form of optical components that were useful for various things, like sensing and thermal scanning. Lys could no longer remember exactly what the initial attraction had been. "But we started playing with them, and we did many of the things that people aren't supposed to do with LEDs—we ran them really hard, we pulsed them at crazy frequencies, and tried to get them to do all sorts of strange things."

It was a classic case of the playful, poke-it-and-see-what-happens mentality of the computer hacker, a category into which Lys clearly fitted. At one point during what his CV drolly described as his "seventeen years" as a PhD student at CMU, Lys had built a system with seventeen disk drives hooked up to his computer, the sort of goofy thing he would do just to see if it could be done. Lys's other great love was designing circuits. But by the early 1990s, the impetus to play was no longer as attractive as it had been in the early days. If you needed more computing power, CMU had a network of five thousand computers. If you designed a nifty little circuit, you didn't actually have to build a chip; you gave it to the computer-aided-design vultures and it would disappear into their automated circuit generation libraries, never to be seen again.

So seeking further amusement, Lys and his friend Mueller began working with LEDs, making lightbar displays. These relied on a phenomenon known as "persistence of vision." You scrolled information across a display that was four feet high but just a few columns of LEDs wide. The eye would retain enough of the information to form an image of the entire picture. Mueller decided that lightbar displays had commercial potential for advertising purposes. The pair formed a company, Stone Age Technologies ("Because there weren't supposed to be any technologies in the Stone Age, right?"), to market them. The displays used thousands of LEDs, red and "that awful gallium phosphide green." But lightbar displays were not a commercial success. "We sold a few of them, but there were always two issues that came up," Lys recalled. "The first was, Can you make the image sit there and hang? And of course that's fundamentally impossible with this kind of thing. The other question was, Why does it cost so much? And the answer is, Because the product you really want has already been invented, and it's called the television. It took us several years to realize that you just couldn't convince people that they wanted this thing." By 1996 their first attempt at a company was moribund.

But at least the lightbar display served to gain Lys and Mueller entry to the lighting industry and to give them an awareness of the unexpected opportunities there. Checking out an entertainment industry trade show, the pair noticed what seemed to them a giant hole in the market. Namely, that while the high-end, super-bright segment of professional lighting seemed well supplied with product, at the low end, below 200 watts, there was almost nothing except garden-variety halogen light sources. If you needed different colors, you bought a couple of these lights, stuck filters in front, and hooked them up to your dimmer rack. "We knew that there was something to be had there. Of course it didn't take us very long to trip over LEDs as being the right way to do that, because we had piles of them sitting around doing nothing. And that's really how the idea got started."

Right around this time, Lys got his hands on some of Nichia's first bright blue LEDs. "They were six dollars apiece, but they were cool. . . . The other thing that Nichia's salespeople were pushing was pure green. Originally, I was like, They're six dollars apiece, there's no way, I'll just have to use the other greens. The sales guy was like, I'm going to send you some of the pure greens, you have to play with those. And they were fairly impressive. . . . One of the initial thoughts George and I had was that this was one of these things that would probably be aided by the network effect. That is, whereas one unit probably wouldn't be very impressive; but a hundred of them, now that might be really cool. And I don't mean a hundred LEDs, I mean a hundred *clusters* of LEDs.

"The interesting thing was that, at that point, there weren't too many other people in that space. I mean, nobody was working on LEDs. It was kind of like this forgotten technology that you used for indicators, they were seven cents apiece, and that was that. So we had piles of LEDs from robotics, and George would say, Drive them harder, try this, try that, try something else, y'know, just keep working on it. He was very much the sort of person who would say, Oh no—you can do better than that, I know you can do better." Mueller would keep challenging Lys, betting him five beers that he couldn't make a microprocessor-controlled light fixture based on LEDs, or find a way to address the individual lights using Ethernet protocols. Time after time, Lys would rise to the challenge.

* * *

I meet this odd couple, Lys and Mueller, in Color Kinetics' headquarters. These are unexpectedly located, not out in some anonymous high-tech business park, but smack in the middle of historic downtown Boston, directly across from Ben Franklin's birthplace, just behind the Old South Meeting House where the uprising that led to Boston Tea Party began. An appropriate setting, perhaps, for a company whose goal is to revolutionize the lighting industry.

Lys the technological wizard is relatively compact in build with a chubby, square-shaped face. A deceptively sleepy look conceals a sharp intellect and a dry wit. The child of Ukrainian immigrant parents—Ihor is the Ukrainian version of Igor ("The Russians don't have the *h* sound so they have to use a *g*")—he was born in 1969 just outside Washington, DC, where his father was a hydrological engineer on the Potomac River.

Mueller is tall, slim, and handsome, with curly longish light brown hair and a winning smile. A self-confessed lover of excitement, he evinces an infectious, almost puppy-dog-like enthusiasm. Born in 1970, he grew up in Bloomfield Hills, Michigan. This is an affluent suburb of Detroit, but growing up in a divorced family taught him self-reliance. As a kid he did all sorts of things around the house, from sewing on buttons to fixing lawnmowers. In essence, Mueller is about as close as it gets to your archetypal, all-American entrepreneur. He revels in the sheer joy of creating something new. Stone Age Technologies was his first company, started while he was still at business school, doing a degree course he never finished, as a way of putting what he was learning into practice. "Business school doesn't teach business," he grins. "You've just got to roll up your sleeves and do it." Lys adds that, "George is a big proponent of, when you've got something—try and sell it."

But it was on his second start-up, Internet Securities, that Mueller really cut his teeth. This is an online service that provides news and data about the world's emerging markets. Mueller formed the firm in 1994 with his brother Gary, who credits him with having had the foresight to realize the significance of the Internet as a medium for information delivery. Mueller flew to nineteen emerging markets in Eastern Europe, Latin America, and Asia. With him he carried crammed into four suit-cases—the maximum allowed—enough equipment to set up an entire office. In each location, he built a remote Internet site based on the best available technologies of the time, like Cisco routers. He signed up infor-mation suppliers, then found customers willing to pay for the data. The company survived the collapse of the dot-com bubble and continues to flourish to this day. In 1999 the brothers sold 80 percent of the company for $43 million.

Long before income from Internet Securities began to flow, however, the thrill of creating something from nothing had gone. By 1996 Mueller was ready to move on to his next new thing. He started Color Kinetics based on his savings of sixteen thousand dollars, plus forty-four thousand dollars' worth of credit card debt. (These days, when Mueller delivers lectures in business schools, he tells students that number one on his list of things that people who want to start their own businesses should know is the maxim: "Credit card companies do not read business plans.") In addition, Color Kinetics raised an initial round of financing from angel investors, "what a lot of people call family, friends, and fools."

Having had the original insight that computer-controlled solid-state lighting systems were viable, Mueller and Lys began building prototypes in the latter's two-bedroom apartment in Baltimore. Color Kinetics' first big break came at the Lighting Dimensions International show in September 1997. Rounding up four interns from MIT's Sloan School of Management to assist them, Mueller and Lys stuffed their prototypes into backpacks and flew to Las Vegas. There, they bought some carpet remnants, a card table, and a laser printer—"You want a brochure? We'll print one out for you!"—and pitched their tent amid all the high-tech booths at the show. The makeshift nature of their materials did not seem to matter. Lighting designers took an immediate shine to Color Kinetics' novel marriage of LED lighting fixtures and computer smarts. They began ordering product on the spot, before the young entrepreneurs even had time to determine pricing. "It was almost as if they wanted to hug us," Mueller recalled. The pair walked away with a full order book and an award for architectural lighting product of the year.

To understand the attraction, it helps to know something about the conventional technologies that then held sway in the professional lighting market. Most derived from theater lighting. Your basic theatrical light is an unglamorous object known as a PAR can,* a simple metal case housing an incandescent bulb. The only control you have over such lights is the ability to dim them. Look above any stage and you will see PAR cans dangling from racks. Color is produced by gels, sheets of translucent material placed in front of the fixture in the path of the beam. These gels

---

* The initials stand for parabolic aluminized reflector.

have a limited life, especially in saturated (i.e., pure) colors. The color fades or the sheet melts, then it has to be replaced.

PAR cans are static and cheap. There are also moving lights, sometimes also known as intelligent fixtures. These are more expensive. They use a built-in stepper motor to manipulate the light in some way. For example, they can change the color of the beam using a glass dichroic filter, or the shape of the beam using a *gobo*—wonderful name—a circular stainless steel plate with holes cut in it. The motors are remotely controlled by signals from a lighting console. With these lamps, there has to be a physical component that moves around, for example, glass filters moving back and forth or a color wheel. But every time you have a part that moves, you have a part that fails. (Especially when your motor-driven light fixtures happen to be located halfway up the exterior wall of a luxury hotel on the edge of a desert, say, subject to sandstorms and constant 40-plus-degree Celsius temperatures.)

Red-green-blue LEDs give control over all aspects of light. That is, the ability to change hue (color), saturation (how much white is present), and brightness (amount of light). These aspects can be electronically modulated with such speed and precision that the changes take place too fast for the eye to comprehend. Color-shifting means you can achieve very subtle changes of hue and millions of different shades. The tiny devices could also be packaged in modules that were much smaller than conventional lights and had no moving parts. Each individual device could be addressed, like computers and printers in a network, making it easy to control enormous numbers of LEDs.

Color Kinetics took the set-in-its-ways professional lighting industry completely by surprise. "No one in this market seemed to understand cutting-edge technology," Mueller said with relish. "You could just come in with current-generation microcontrollers and Ethernet protocols and kick the crap out of these companies at their own game. It was a threefold attack: try and bring significant high-tech brainpower into a low-tech industry that's kind of old hat. Along with that, bring in this brand-new disruptive technology and use that as a source for attacking the market. And the third part is, bring in the aggressive high-tech style. I mean, we're not a staid lighting company, we don't grow at 4 percent a year.

We're not happy with that, we want to grow at a blistering pace, we want to get out there and sell. And we want people who are from the tech mindset: if you don't run, you're going to fall behind, or you're going to get eaten by a bigger player."

In 1998, following the success at Lighting Dimensions International show, the company began hiring staff. Their first recruits were not, as at so many tech-heavy engineering school spin-offs, fellow geeks. David Johnson, Color Kinetics' first employee, was a finance guy. "Initially my wife was a little skeptical when I told her I was going to work for a company with no revenues that made flashing lights, strobes, and color washes," Johnson told *Metropolis* magazine. "She asked me if I was betting that disco was coming back."

Another early hire was Kathy Pattison, who headed up marketing at the fledgling firm. The seven-year Apple veteran was immediately struck by Mueller. "He had a ponytail, and he oozed enthusiasm, panache, and charisma. The day I met George I thought, Oh my God, this is going to happen—it was a clean slate, a new company, technology, category, concept." At Apple, Pattison had seen the personal computer completely change the basis on which things were done. The historical analogy with the PC was not lost on Mueller himself, who is not by nature a modest man. "We are at the exact same point in time as the early computer industry," he told a group of MIT students in 2003. "To me, looking at where we are, it's like looking at Jobs or Gates."

Early on, Color Kinetics was advised on business strategy by one of its individual investors, Noubar Afeyan, a biotech entrepreneur who had founded several successful start-ups. "He looked at our product and said, Make sure your intellectual property is solid," Mueller recalled. "We started looking at the intellectual property in this space, and lo and behold—there wasn't any! I mean, it was a greenfield opportunity." To be sure, there were a lot of LED patents, there were some display patents, but there were almost no solid-state lighting *systems* patents. "We were amazed, I mean, absolutely floored. So we engaged multiple patent counsel to help us out and just prosecuted like crazy—invented, invented, invented on everywhere we thought this technology would go, and everywhere we could bring it."

Color Kinetics filed its first patent, "Multicolored LED lighting method and apparatus," in August 1997. By mid-2005, an accommodating US Patent Office had granted the company 38 patents, with more than 130 applications pending. This audacious intellectual-property grab would subsequently cause enormous ructions within the nascent solid-state lighting industry, especially since some of Color Kinetics' patents were extremely broad in nature, while others—such as one describing a color-changing underwater light—seemed to many to be obvious and lacking an inventive step. Color Kinetics was prepared to license its technology, but also to sue the pants off anyone it suspected of infringing. In 2002 a legal battle began between the company and a group of its rivals. At the time of writing, it had yet to be resolved.

Getting funding would prove more difficult for Color Kinetics than staking out intellectual property. In 1999, to assemble a further round of financing, Mueller attempted to shop the company to venture capitalists. But the moneymen did not want to know. Back then, the madness of the dot-com bubble was raging: cashed-up venture capitalists were throwing money around like drunken sailors. "We would continually hear, Are you guys a dot-com company? Are you a B2C company? How about a B2B?" Mueller scoffed. "It's incredible how unimaginative most VCs are. Once someone does a pet-food-on-the-Internet business, they all want to do pet-food-on-the-Internet, they're like sheep—or lemmings. And we were like, Enough! We're not Internet, or telecom, or healthcare. We're not any of those. We're a solid-state lighting company and we've got this great patent portfolio."

Mueller and company found that it was more difficult to try to educate venture capitalists on their model and its value than it was to get private equity, either from angel investors or from strategic investors like LED chipmaker Cree, which came on board in 2001. Only just before Color Kinetics went public in April 2002 would venture capitalists wake up to the realization that here was a real company, with real customers, real revenues, and real profits.

*   *   *

Back in 1997, winning the award at the Lighting Dimensions International show was confirmation that Color Kinetics was onto something. Initially, however, they didn't quite know what. Mueller's original idea was to target the theatrical market, which obviously used colored light, the strong suit of LEDs. Perversely, however, the reaction from that market was negative: You're far too expensive and not nearly bright enough.

Over succeeding years, brightness and price would become less of a problem. In the same way as the microchip industry follows Moore's law, so the solid-state industry conforms to Haitz's law. First formulated in 1999 by Roland Haitz, a former research director at Hewlett-Packard (later Agilent), this law states that every decade in the forty-plus years since their invention, LEDs have increased twenty-fold in terms of their light output. During the same period, the cost of the devices has fallen to one-tenth.

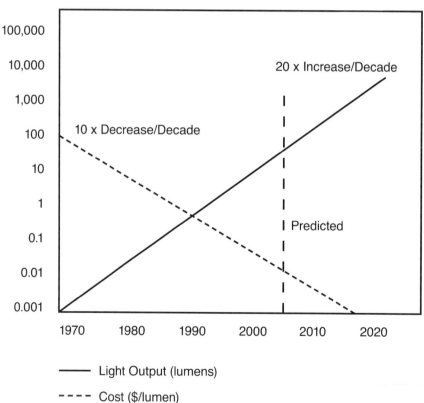

HAITZ'S LAW

The first major applications of Color Kinetics' revolutionary technology would come outside theaters, not in them. For example, the Loews movie chain wanted a new marquee sign for its flagship theater on New York's 42nd Street at Times Square. The sign, a blade that would stand six stories tall by eight feet wide, was due to make its debut on New Year's Eve 1999, when hundreds of thousands of people would crowd into the square for the famous countdown to the ball drop. Brett Anderson, the lighting designer in charge of the project, heard about Color Kinetics in early 1998, by chance, when a colleague in Edinburgh faxed him for information about the fledgling firm, which at the time had all of four employees. Anderson contacted Mueller, who sent him some sample modules. The designer was looking for something that would yield the maximum number of effects. He thought that, at best, LEDs would enable him to change color en bloc as well as dimming and brightening. What he got was the ability to control each of the quarter-million-odd lighting elements individually, so that the whole sign could be sequenced, inch by inch. It could be programmed to shimmer with waves of intense color that seamlessly morphed from one hue to another. These color sweeps could be seen from six blocks away. And, unlike with traditional fixtures, there was no jerkiness as the sign stepped from one color to the next.

Within a few years, theaters, buildings, bridges, and monuments all over the world would be similarly iridescing like squid, cycling through the colors of the rainbow, flashing and dancing with dazzling light. Color-shifting the LED way had come to stay.

*   *   *

To get a better handle on what was happening, I was keen to meet a specialist in lighting design, a profession I had not previously encountered. I asked Fred Oberkircher, who has been teaching lighting courses at Texas Christian University in Fort Worth for twenty-five years. He recommended Jonathan Spiers, Europe's top lighting designer, whom he described as having "the most exquisite color sense I have ever seen." As it happened, Spiers was the person who had contacted Brett Anderson asking for information about Color Kinetics. I traveled to Edinburgh to visit him at his studio in the Dean Village.

The Dean Village is a little community on the Water of Leith, the river that runs through Edinburgh. It is separated from the rest of Scotland's capital by virtue of being located at the bottom of a ravine. Having been to school in Edinburgh and having had a friend who lived in the Dean Village, I thought I knew this area quite well. But I was not prepared for Well Court Hall, Spiers's studio, which has to be the most extraordinary set of working premises I have ever come across. The hall is located in a mini clock tower, up several flights of stone-flagged stairs. Featuring wooden hammer beams, stained glass windows, an enormous fireplace, and a minstrel gallery, it was built in 1884 by a philanthropist as a communal gathering place where local working men could meet and talk, without the influence of either religion or alcohol.

Jonathan Spiers has been lighting the insides and outsides of buildings for more than twenty years. He is an affable, quiet-spoken man with a slightly bemused air and a self-deprecatory sense of humor. I begin by asking him to explain, what exactly is a lighting designer? "We're a bunch of misfits," he replies. "If you lined up all the lighting designers along a wall, you'd think, Who the hell are these weirdoes?" Though a professionally trained architect with a long line of letters after his name to prove it, Spiers counts himself fortunate to have discovered school theater when he was twelve years old. He became hooked on lighting. Other lighting designers, like Spiers's friend and colleague Paul Gregory, the founder of New York–based Focus Lighting (the company for which Brett Anderson works), come to architecture from a theatrical background. Spiers and Gregory often collaborate, teaching classes that climax with "guerilla lighting." That is, "attacking" buildings with light, unbeknownst to their owners, to show students how lighting can be deployed to maximum aesthetic effect.

You might think that architects would plan for the nocturnal appearance of their buildings—it is, after all, dark half the time—but typically they do not. "Architects gave away the lighting design of their buildings years ago, to the electrical engineers," Spiers says. "There should not be a profession called lighting designers. When the issue of how does it look in the nighttime arises, architects say, Oh, the engineer does that. But that's bullshit!" It seems education is to blame. "Lighting design at architectural colleges has tended to be a matter of formulas: calculate the number of amount of fluorescent tubes needed to illuminate a space x meters by y. It's the wrong way to inspire people about what amazing stuff light is."

This unwillingness of architects to take lighting seriously is an old problem. In the early twentieth century, with the coming of modernism, there was considerable enthusiasm, especially in the United States, for the use of electric light as a new building material to create what Dietrich Neumann, in his magisterial *Architecture of the Night*, describes as "a future luminous architecture." In American cities during the 1920s and 1930s, thousands of buildings were lit up. Skyscrapers were ideal subjects. Neumann quotes a 1925 article from the *New York Times* that describes the Manhattan skyline as "a fairyland of night . . . a huge city of illuminated castles in the air."

Unfortunately, however, much of the impetus for picking out the outlines of buildings with strings of incandescent bulbs and, subsequently, for lighting up entire sections of buildings with white or colored floodlights, came from companies that sold electric power. They saw large-scale illumination as a good way of generating revenue, especially during the 1930s, to make up for industrial customers lost during the Depression. This made architects leery, their suspicions being compounded by the fact that most lighting designers came from a theatrical background and were thus ipso facto infra dig. Architectural journals hardly deigned to notice the subject of lighting design. Architects exhibited what the trade magazine *Light* called "a peculiar reluctance to be educated." For most, it was all a little too much like Las Vegas.

The ephemeral art of architectural illumination has also suffered from external factors, followed by collective amnesia—lighting plans were seldom written down—necessitating its subsequent reinvention. The blackouts of World War II put a stop to most outdoor illumination. It was not until 1964 that floodlighting of the top thirty stories of the Empire State Building began. The oil crises of the 1970s caused a second hiatus. Floodlighting of public monuments did not resume until the 1980s. The coming of computers enabled a move from static displays to programmable lighting sequences. But until the advent of LEDs, maintenance issues often proved crippling for designers trying to use light in innovative ways.

It has thus taken lighting designers a long time to achieve legitimacy, to belatedly respond to the earlier utopian visions of a new luminous architecture. Their involvement in architectural design continues to be the exception rather than the rule. The conflict continues: lighting designers like to do exotic things; architects are forever complaining that such schemes do not reflect their design goals. The lighting of buildings is often still merely an afterthought rather than something integral to the structure.

Neumann points out that "light frescoes" had been predicted in the 1920s. But it was not until the coming of large-scale LED displays like the NASDAQ screen in Times Square that such visions could be realized. Even then, the challenge remains of how to integrate the screens with the architecture upon which they are hung. Recently, however, lighting designers have begun taking advantage of the unprecedented controllability of LEDs. Here are some contemporary examples, the first one by Jonathan Spiers.

"We proposed this idea for a big financial center project in Dubai. It was the gateway to an entire site of about fifty buildings, a nine-story building, with two office blocks on the sides and a three-story slab over the top, and you walked in underneath. Dubai started life as a trading nation, buying and selling pearls and spices, then they found all this black stuff [i.e., oil] under the ground and made a lot of money from it. But a financial center is all about buying and selling, doing a deal, a transaction, which involves interaction and communication. So I asked myself, What would make people want to walk underneath this gateway?

"I had this idea that we'd have a grid of LEDs embedded in the ground, RGB, Color Kinetics or whatever, and nine stories up, a TV camera. This is not cutting-edge stuff, it's using existing video processing technology, with a little bit of a twist. Basically, you'd walk onto the floor from one end, the camera would pick you up, and you'd get an aura from the LED units around you changing color. So from a background color of blue, it might go to cyan. Someone else walking from another direction would have the same color, and as you meet, maybe it goes to pink, or a more saturated color. If three people get together, it goes to red, and if six people, to yellow. And this would all be happening live, in real time. One person splits off because he's got to go to a meeting, he takes his aura with him, he's back to his cyan color, and the collective aura of the group diminishes, because there's more power in collective debate and discussion. Imagine it, it'd be great—there'd be little kids rushing around going, Look, I've changed color! You couldn't do that with any other source than an LED in terms of that fast response, low power consumption and all that."

Unfortunately, at the last minute, much to his chagrin, the client backed out of implementing this design, but Spiers is determined to use it on another project. Other designers have been more fortunate. For example, James Carpenter, who designed interactive walls for the south and the north facades of the new 7 World Trade Center building in New York. The function of the facades is to conceal an electricity substation that occupies the building's first six stories. At dusk, 220,000 blue and white LEDs illuminate the walls from within. Motion-sensing cameras mounted high on building corners track pedestrian movement. As people pass on the sidewalk below, a multistory strip of vertical cobalt-blue light gracefully follows them, making the wall a kind of light show for each passerby. When many pedestrians walk by simultaneously, they create complex patterns of moving columns of light. "It's not Times Square," Carpenter told *Business Week*. "We wanted to be very subtle and not too bright, so it illuminates quietly and uniformly."

Responsiveness to passing traffic is also a feature of an LED installation near where I live in Melbourne. It is a three-hundred-meter-long wall located just before the junction of a bypass and a ring road. Functionally, this wall is an acoustic barrier. Aesthetically, it serves an entirely different purpose. Embedded in the transparent screen-printed acrylic panels that make up the wall are 935 custom-made weatherproof light fittings called luminaires. Stacked in columns of five, one meter apart, each luminaire contains a cluster of nine LEDs. The LEDs can be individually programmed to produce any color. The display is triggered by transducers in the road and controlled by computer. As you drive past, the patterns change constantly, streams of light rippling across the wall. They vary depending on traffic density: the more vehicles on the road, the slower the rate of change (so as not to distract drivers).

The original idea, according to Robert Owen, the artist who designed the wall, was to suggest the flickering television sets that can be seen through the lace curtains of the houses that belong to the mostly Italian migrants who live nearby. But the LED matrix can also function as giant low-resolution video screen, a blank canvas that may be offered to other artists at festival times, or as a bulletin board that can send text messages. ("Are we there yet?" was one that waggish designers ran during tests before the bypass opened.)

*   *   *

Few people have spent more time thinking about the implications of LEDs than Sheila Kennedy, a Boston-based architect who teaches at Harvard. She sees the new solid-state lighting as very different than previous forms of lighting such as candles and incandescent bulbs. "The first point to make is that LED illumination is cool light in terms of device temperature. The efficacy of the semiconductor is so great that a vast percentage of the energy is transmitted into light with very little loss in heat." In fact, Kennedy sees the way LEDs generate light as being more like the bioluminescence produced by living creatures such as fireflies, bacteria, and fish. "From cool light and the idea of nature, suddenly you can imagine that there's a new level of integration that's possible, and that changes everything. Light can be introduced and blended into a variety of different hosts, materials that have properties that are entirely different from the way our culture thinks about light, which is as a kind of object fixture."

At Harvard's new school of film and digital video, Kennedy has integrated LEDs into a curtain wall made of fabric that automatically brightens and dims in a series of behaviors that mimic cinematic light. She has also been working with textile manufacturers to produce fabrics with LEDs embedded within them that can be used in a variety of innovative ways. For example, a blanket that would generate enough light to read by. "What we're beginning to understand culturally is that light is a material, an emissive material. . . . I think you'll see the next generation of architects and artists become more comfortable with these materials and try to understand their properties."

One such artist is James Clar, twenty-six, who has designed an "energy mesh" that will wrap around the Habitat Hotel, to be built in Barcelona in 2008. This takes advantage of the affinity between solar cells and LEDs, the former generating electricity in the form of direct current, on which the latter runs. The mesh incorporates five hundred RGB LED nodes. During the day, the nodes will collect energy from sunlight. At night, the mesh will be programmed to glow in specific color schemes whose brightness is determined by embedded photosensors. Thus the glow that envelops the building will change dynamically, depending on the weather and the season.

＊　＊　＊

Artists have always been fascinated by light. Think of Caravaggio, who used *chiaroscuro* (Italian for "light-dark") to transfix his subjects in dramatic shafts of light against somber backgrounds. Or the Impressionists, for whom light was the central concern, as with Monet, and his light-drenched haystacks, or Degas, and his lime-lit dancers. Then, in 1963, a Brooklyn-born artist named Dan Flavin created a new medium out of light itself. He hung an industrial-grade yellow fluorescent tube on the wall of his studio at an angle of forty-five degrees, cheekily titling his priapic work *The Diagonal of Personal Ecstasy*. From there, Flavin went on to produce a whole series of light installations, each one made up of fluorescent tubes in basic colors arranged in rectangles, parallelograms, and other simple geometrical shapes. In so doing, according to *New York Times* critic Michael Kimmelman, Flavin "consciously blurred the distinction between art and architecture, seizing architecture as part of art's sculptural vocabulary, incorporating corners, walls, doorways, and windows."

There were of course limits to what could be done with linear fluorescent tubes. Other artists have sought to push further what can be achieved through the use of light. Towering above them is James Turrell, who has spent his forty-year career learning how to shape light, exploiting its physicality, what Turrell calls its "thing-ness." He has created light installations that appear so substantial that, believing them to be solid, viewers have been known to lean on them, fall over, break their wrists, and—this being America—sue the artist for damages.

Turrell is best known for Roden Crater, an extinct volcano located near Flagstaff, Arizona. There, for many years, he has been boring into the rock to create observation portals through which the sky and other celestial phenomena may be viewed. To fund this magnum opus, Turrell undertakes commissions, including lighting design for buildings. During the course of this work, he was introduced to LEDs. Turrell has said recently that he wishes LEDs, and the precise control over light they offer, had been available to him decades ago when he began working with light. "I thought [the technology] would come a lot earlier; but here it is, so I'm not going to complain."

Turrell has used Color Kinetics LED fixtures in several recent installations. One such is glowing blue light chambers for the lobby and the library at Greenwich Academy, a private girls school in Connecticut. Another is a black granite swimming pool, in the basement of a barn, also in Connecticut, owned by property developer Richard Baker. The pool is surrounded by thin ribbons of light that morph from flaming reds and pinks to cool shades of blue and purple. Turrell's largest LED project thus far is the Takarazuka School of Art in Osaka, a glass-fronted tower block designed by Tadao Ando, Japan's leading architect. Throughout the night, the facade of the building changes color. It begins with one hue, which very gradually changes to another.

In many ways light artists like Turrell and lighting designers like Spiers have only just begun to investigate what the new technology has to offer. "One thing that's going to happen," Spiers predicts, "is that people will start to play. I've got little color-changing LED things that I gave my kids. And they sit there in the dark playing with these things. The guys on my team are all basically as mad as I am, they're interested in the medium, and we're always trying to push our clients into doing something interesting." Much of what is currently being done is not so interesting, color-shifting simply for the sake of it. Gaudy me-too buildings are popping up everywhere, nowhere more noticeably than in mainland China, on the illuminated new skylines of Beijing, Shanghai, and a host of other Chinese megacities.

*　　*　　*

Despite all their advantages for the lighting designer, LEDs also have at least one big disadvantage, one that paradoxically derives from the rate at which the technology is improving. "Let's say a client wants something really zizzy [*sic*]," Spiers explains. "You specify five hundred of these LED fixtures to wrap around the facades of a shopping mall, and you program up something really cool, so there's these subtle color shifts and all the rest of it. The issue is, after three or four years, there's a fault in five of the fixtures, for whatever the reason might be—maybe water gets in, maybe they blow up, maybe vandals come along and trash them. So you need to go and buy another five. You call Color Kinetics and ask them to send five of those Color Blast fixtures that you bought four years ago. They say, Oh, we don't make those anymore. You ask, Why not? They tell you, Because the LEDs we've got now are far better than the ones we had four years ago: better colors, more purity, and they last much longer. And you say, No, no, no—I've got 495 other ones, I want replacements that are exactly the same! In that sort of situation, it's not good to be an early adopter." As another designer points out acidly, "We call them light *fixtures* for a reason."

The blistering rate of change in the solid-state lighting industry contrasts markedly with the sedate pace of the construction industry. "The lighting business is a lot slower than we thought," Ihor Lys complains. "Part of that is what we call the 'spec cycle.' For there to be a demand for lights, buildings have to get built, and that's a multi-year process. We had assumed that they do just-in-time engineering on a lot of this stuff, and to some extent they do. The difficulty is that schedules slip, so all sorts of things happen, where the lighting plan for a building has been designed and fixed and waiting for the building to be built for five years. That happens quite regularly, actually. But for someone trying to start a company, that's a terrible set of conditions. You have to try and convince people to change plans, or you have to submit your stuff, they take a look at it, then they tell you that you don't have enough installations for them to trust what you're telling them. Or, that they can't put you on such-and-such a project because the bids have already gone in, but they have another project that's going to be built four years from now, and would you like a piece of that?"

Despite such frustrations, by 2005 Color Kinetics was doing "very nicely, thank you." In just its eighth year in business, the company had racked up $53 million in annual sales, of which $4.3 million was profit. Color Kinetics lighting modules were now being used in some very large-scale projects, like the Hard Rock Hotel and Casino in Las Vegas. "These are LEDs lighting up an eleven-story building," George Mueller bragged. "Who on earth would think that these little indicators which used to be in clock radios or power-on buttons would be lighting up a whole building that's like a couple of football fields in area? Ten years ago, you couldn't imagine that LEDs would be lighting something as big as this, doing a great job, beating out the traditional technology and saving a significant amount of operational cost. I mean, this hotel is saving approximately $40,000 a year on power and maintenance costs; they went from 44 kilowatts to 4 kilowatts. It's just a phenomenal story."

In 2004 Color Kinetics announced that it was moving on from color with a series of white light products targeted at the professional lighting designer. At the time of this writing, these products were already in their third generation. The efficiency of their light sources was 40 lumens per watt and rising. "Within the next few years we're going to be shipping product in the 70 to 100 lumens per watt range," Mueller predicts. "As soon as we hit 70 lumens [the point at which LEDs become competitive in light output with compact fluorescent lamps], it starts to get very exciting; we get to a 100 [competitive with linear fluorescent tubes], and it's dancing-in-the-streets-naked kind of exciting. So it's like the rocket boosters are going." He brings his hands together to form an inverted V and vibrates them. "Everything in the industry's rumbling, and we're just waiting for it to blast off, right?"

*   *   *

Thus far, we have looked at the aesthetic benefits that the unprecedented control over light sources brings. But there is also another, related set of considerations, the enormous implications of which are only beginning to be understood. "The other thing that is going to happen that we have no concept of right now," Mueller explains, "is that we are going to start learning a lot more about human factors and human behavior under ambient lighting conditions."

We already know that light influences all sorts of things, most notably human health. Studies of maternity wards have shown that whether the lights are left on at night has a slight influence on the growth rate of newborn babies. Lighting also plays a major part in what constitutes a healthy environment. Hospital patients lying immobilized in windowless wards are often forced to look at the ceiling and walls under the same lighting conditions for hours if not days. LED lighting systems would make it possible to change the color temperature from warm—reddish—to cool—bluish, replicating lighting conditions outside. Such systems may even be used to treat those who suffer from, for example, seasonal affective disorder (SAD), aka the winter blues. This is thought to be the result of a biochemical imbalance caused by the shortening of daylight hours and the lack of sunlight in winter.

Schools that use daylight and view-windows report dramatic increases in student performance compared to those confined to electrically lighted, enclosed classrooms. In the office, too, it is known that there is a correlation between worker happiness—and more important, from an employer's point of view, productivity—and the type of lighting under which they work. "If you give people a choice of what kind of office they want," Mueller says, "they typically say a corner office with windows, and if they can't have that, they want daylight. I think, in twenty years from now, you might see lighting mimicking daylight outside. It might become common to use light to get your biorhythms into the cycle of the day and the season of the moment; if it's fall, it might be warmer color temperatures and a shorter daylight cycle, and so on."

The ability to control color temperature is also of great interest to designers of next-generation airliners, who are planning to switch to LED systems en masse. Indeed, it is in the air that some of the most sophisticated applications of solid-state lighting may initially occur. The prospect of ultra-long-haul—nineteen-plus-hour—flights is forcing the airline industry to rethink the way interior space is lit. For example, in the cabin of the Boeing 787 Dreamliner, due to enter service in 2008, harsh fluorescents will be replaced by an LED system that can be programmed to replicate the day-night cycle. At mealtimes, a subtle red blush will be added in an attempt to make airline food look more appetizing. After dinner, LEDs on the ceiling will twinkle like stars, mimicking the sky at night. Next morning, the cabin crew can bring up the lights gradually, to create an illusion of sunrise.

George Mueller and Ihor Lys were not the only ones to begin tapping the potential of solid-state lighting. The next chapter investigates two other, very different entrepreneurs, the companies they formed, and the uses they found for the new high-brightness LEDs.

CHAPTER 8

Dave Green had a favorite routine that he used in presentations to impress potential investors in his new company, Carmanah Technologies. The company's first products were navigation lights made from a combination of LEDs, solar cells, and batteries. Green would ask the biggest person in the audience to come forward and invite him to whack one with a hammer. Encapsulated in hard plastic, the lights were virtually unbreakable, which of course is an unusual thing for a light to be. The idea was to get everyone in the room talking, to focus their attention. Green had pulled off this stunt many times, and it had always worked like a charm. One particular night, however, in front of about fifty or sixty locals at the Union Club of British Columbia in Green's hometown of Victoria, things went awry.

"I got this big hulking guy to come forward," Green recalled. "I gave him the hammer and I said, Here—bet you can't break this, it can withstand anything. He put the light on the floor and he whacked it, and this thing flew two or three feet into the air, and when it landed, he whacked it again, and it flew up again. I said, OK, that's enough. I took the light and I put my hand over it—it has to be dark before it'll light up—to show the audience. Well, the damn guy had broken it. The batteries had come right out, they weren't secure, I could feel them moving around inside. So I went behind the screen where I was doing my PowerPoint presentation and I got another one. Anyway, he hit this other light and he broke it, too. And by then everyone was roaring with laughter."

To make matters worse, the Canadian Broadcasting Corporation was filming this debacle, as part of a series on entrepreneurs. The program was broadcast repeatedly. For years afterward, when Green met people, they would ask him what he did, and he would tell them that he made solar-powered lights for navigation. "Invariably they would reply, Oh, I watched a show on CBC about that. There was this guy, he broke his lights in front of potential investors. And I would go, Yeah, yeah, that was me. It haunted me for years."

Nonetheless, Dave Green would have the last laugh. By 2005 Carmanah would be the world's leading supplier of solar cell-based LED products. In that year the company had sales of almost $40 million, more than a hundred employees, and an annual growth rate of around 70 percent, making it one of Canada's fastest growing firms.

*   *   *

Like Cree's Neal Hunter and Color Kinetics' George Mueller, Dave Green is a serial entrepreneur. On graduating from the University of British Columbia with a PhD in oceanography, Green discovered that there were no jobs available in academe. So he and a couple of friends started what blossomed into a group of ventures, all of them initially with a marine flavor. Seastar Optics, for example, began by attempting to make a new type of oxygen sensor for seawater. This involved the use of fiber optics and a laser. It was hard to get the laser to shine down the fiber, so the team had to come up with a way to link the two. The coupler they developed would turn out to be of considerable commercial value for optical communications. But not in the short term.

In 1989, after fourteen years of working his butt off with very little to show for it, fed up with continually having to put his personal finances on the line to keep the underfunded firms afloat, Green decided that he needed a break. He sold his share in the businesses to his partners and went off to the University of the South Pacific, which is based in Fiji. He spent three happy years solving all sorts of problems for the islanders. One of the things that Green encountered down there that he had not previously been exposed to was renewable energy, especially solar panels. He used most of the money he got from selling out to buy himself a small sailboat, yachting being his passion. ("I love the complexity of yachting," he told me, unprompted. "It completely engages me.") After leaving Fiji, he and his wife and three daughters sailed around the South Pacific before heading back to Victoria, which they reached in 1993 after a nine-month voyage.

"By then I'd been out of the country for five years, I was forty-five, I had no job to come back to, and not much money left. When I came back, I found that my ex-partners were now wealthy. They'd sold the laser-fiber connector business to an American company. They had a lot of people working for them, they were doing great, whereas I was almost broke." He started doing consulting work for the forestry industry, dubbing his consultancy Carmanah (pronounced *car MAN ah*, the name of a valley in a local old-growth, state-owned forest.) "But I wanted to get into making products again. So I went to [Canada's] National Research Council to get a small grant—fifteen thousand dollars—to look at the idea of a solar-powered anchor light for boats."

It would be, as Green knew from his own experience, a great thing for sailboats to have self-contained lights that could run independent of batteries. Maritime law mandates that when your boat is moored or at anchor, you must show a white light with a range of two miles. In practice, however, yachties who leave their boats in a certain place for more than a day or two don't usually bother leaving a light on, because where's the power going to come from? Many was the morning out on the ocean that Green had woken up to find his boat battery drained and his navigation lights out as a result.

Solar panels and LEDs are beautifully matched. Indeed, it's as if they were made to go together, the one extracting electric energy from light, the other emitting light from electric energy. The combination of the two technologies would be key in everything that Carmanah did. The problem was that, back in 1994, there were no white LEDs available.

Green thought that yellow ones might do the job, so he and an electronics guy he had hired to help him mocked up a prototype LED-based anchor light. "I put it on my boat, walked around the harbor, but I couldn't see the damn thing for more than about 150 yards. So that wasn't going to work." But at least it was a start. Green and his engineer came up with the idea of a completely self-contained—"potted," to use his word—solution consisting of solar panel, LEDs, plus a storage device known as a supercapacitor. Their idea was to invent a lamp that would make light forever. They proceeded to patent the design. "It's completely potted, so you can't change the batteries. Supercapacitors and solar panels and LEDs last forever, if you treat them right. So if you put one of these lights in, you'd never have to go back to service it. It would last you until you were dead."

But nobody wanted to buy Carmanah's first batch of lights because they were not bright enough. "We tried almost giving them away, for making bicycle trails and things, but it just wasn't going anywhere." Then came an unexpected consulting contract from the US Navy, to perform an environmental assessment on a nearby torpedo testing range. In order to gauge the performance of the torpedoes, the position of everything—submarine, tracking vessels, and target—had to be fixed. This in turn necessitated the use of buoys within the strait where the tests take place. "And they're huge, these buoys, they're hazards to navigation, they have no lighting, but they're never in the same place, so you can't mark them on a chart." Following some nasty near-accidents, there had been complaints from the public about these buoys. "So one of the recommendations of the assessment was, You need to have lights on these things. The navy's response was, Well how are we supposed to do that? I thought, OK—I can solve that!

"We decided that we could make a red light, which was a lot easier than a white light, which didn't exist, because by now the high-brightness red LEDs were coming along nicely. So we produced a light, and that allowed us to expand the business. But red lights aren't terribly useful in the marine environment without a green." Navigation lights indicate the orientation of a vessel, hence the direction in which it is traveling, using red for port and green for starboard. "When we started there was no green, but just after we finished making our red light, the green LEDs from Nichia became available. So now we could say we made navigation lights."

Green showed the new lights to the Canadian Coast Guard. "They were delighted with them. They just couldn't believe these things, because before they'd been paying ten times as much for something ten times as big. But the combination of LEDs plus solar cells made everything much more efficient. Because you no longer had to have a great big buoy, which was needed to contain a great big battery pack, which then needs a great big chain to anchor it, and a great big ship to take it out to be moored. And when the batteries die, the great big ship has to go back out to the mooring place and lift the buoy out of the water to change the batteries."

With Carmanah's new lights, you could use much smaller buoys that did not need a big ship to transport them, and that, once in place, did not need to be serviced. Because the current the LEDs draw is so small, they just use the "float"—enough current to equal the self-discharge of the battery. They don't drain the batteries, which basically stay charged all the time. You could also throw in a microcontroller—which, like solar cells, is a natural match for LED technology—to do various energy-saving tricks. Thus the annual maintenance costs for a navigation marker fall from as much as fifty thousand dollars per year to zero.

Navigation was a tough business to crack. "You put these lights out on a buoy in the middle of nowhere, they get bashed, they get soaked in saltwater. Some of the buoys get destroyed, and the bits end up on a beach somewhere." But Carmanah's lights were up to the challenge. "They've developed this wonderful reputation for being indestructible. The lights are so self-contained, they have no on/off switch; I mean, there are no holes in these things. You use a remote control to turn them on, so there's nowhere to leak. We're now in our seventh year since we started making them, and they're still not coming back."

LEDs are also much better than incandescents at pulsed operation. They turn on and off instantaneously, unlike the old navigation lights, which are filament bulbs and as such suffer from thermal inertia. That is, before they can flash, they must first heat up. They brighten slowly and dim slowly, whereas LEDs come on hard, stay on hard, and go off hard, in classic digital-step fashion. Such abrupt transitions are easier to see, extending the range of the lights. Meanwhile, LEDs keep on getting brighter. "I love it," Green says. "Our raw material just keeps getting better. We don't change anything, just put in new LEDs, and this light that used to have a range of one mile is now a two-mile light." In addition to which, the company is learning tricks to squeeze extra performance out of its LEDs. Lens makers are also adapting their designs to match the new technology. Carmanah's most powerful lights have a range of five miles.

Running a flashing light is hard on a filament—they burn out quickly. Carmanah's competitors had spent years developing special bulbs with skinny filaments down the middle, and lamp changers that, every time the filament burned out, automatically clicked over to another one. Such mechanical solutions were intrinsically expensive and required regular servicing. At one-tenth of the cost and weight, Carmanah's lights have revolutionized the navigation business. "We basically just turned the whole industry on its head," Green says. He reckons that his company's penetration in navigation lights is only one-tenth of the potential market. "I mean, it still hasn't worked its way through the world, but it's happening. It's only a matter of time before all marine installations switch to LEDs."

In the wake of their north-of-the-border counterparts, the US Coast Guard followed suit, making their first purchase of Carmanah navigation lights in March 2000. The coast guards were tough customers, insisting that the lights meet their rigorous standards, performing tests on them for years. Once satisfied, however, they bought in quantity. Their endorsement has provided wonderful cachet for convincing new customers.

In addition to its marine lights division, Carmanah also musters an aviation division that sells runway markers to airport operators. "They look a lot like the marine lights, but the optics are slightly different because they have to shine the light up." One recent customer is the US military. "We've sold an awful lot to them for use in Iraq. The reason is they want to set up these bases, but they don't want to put in the permanent infrastructure because they don't plan on being there forever. You just put these lights down the side of the runway, and away you go. But we're also doing real airports in the US." Here, too, the logic of LEDs is persuasive. They last longer, more than fifty thousand hours, versus three thousand hours for a halogen bulb. This ensures low-maintenance costs and better traffic flows, since airport managers are not constantly having to close taxiways to change burnt-out lamps. And LEDs use a fraction of the power, reducing an airport's energy bill (and greenhouse gas emissions). Estimates put savings for operators at 40 percent over ten years.

* * *

To meet Dave Green in Victoria, the pretty, pocket-sized capital of British Columbia, I take the clipper up from Seattle to Vancouver Island, a journey of almost three hours. As the boat noses its way into the little harbor, I note the incongruously massive state parliament buildings. These are fronted by a statue of the queen-empress and a towering wooden totem pole, the only indication of the presence of indigenous people, whom Canadians call "first nations." At the far end of the harbor, the eye is drawn to an extraordinary asymmetrically shaped bridge. The light blue structure is dominated by a massive concrete counterweight at one end that enables the bridge to be raised and lowered in order to let masted boats pass underneath. Built in 1920, it was designed by Joseph Strauss, who would subsequently go on to design a somewhat more famous bridge just north of San Francisco.

I encounter Green at his office, above a cigar store across the way from the bridge. It is a cold, wet, windy, day in early April. "The good news is, this is about as bad as it gets," he grins reassuringly. Green is tall and thin with longish windblown graying brown hair. Casually dressed, he looks every inch a yachtie. In fact, he comes from a nautical background: his father was a naval officer. It may just be because this is Canada, but his angular weather-beaten features remind me slightly of Neil Young. He has a lopsided smile and a shy, unassuming manner.

I mention the seeming mismatch of a high-tech start-up being located in Victoria (population: 335,000). This largely tourist town trades on its British Empire heritage, to the extent of even having a daily newspaper called the *Times Colonist*. "Victoria is actually a very dynamic place,"

Green corrects me, "but in a small way, not a large way. You get lots of smart people here, because they want to live here and they're willing to move. There's a lot of innovation, artistic and intellectual energy, but it hasn't produced any large companies—it's a bit remote from the sources of finance—almost by choice, because I don't think people want to start something big. They just want to do their own thing. But under the surface, there's a lot of very clever small companies here, leading-edge firms."

I ask Green what motivates him to start up companies. "I've always been really competitive," he replies, "always wanted to accomplish things, but not focused around money. That's certainly not been the foundation for choosing to go here or there—you don't go into oceanography if you want to make money! So it's been more just following the interest, and the excitement of doing companies; y'know, when you're scientifically oriented or engineering oriented, it's very exciting creating things, building products, and it's great fun." These days, however, he is more comfortable working behind the scenes as chairman of Carmanah. He has passed the role of CEO to Art Aylesworth who, coming from a sales background, is better equipped to function as the company's frontman. This arrangement also gives Green more time to indulge in his passion, yachting.

Green shows me round the company's manufacturing facilities. These are located across the bridge from his office in a bunch of light blue weatherboard shacks that used to belong to the coast guard. The impression is of a large-scale cottage industry. But this is misleading because from this humble location, as Green proudly—and somewhat bemusedly—points out, they produce products worth millions of dollars a month. (Since my visit, the company has moved to a new manufacturing plant nearby.)

Initially, finding funding for Carmanah was hard. Local venture capitalists didn't want to hear about anyone who had made an actual product. "They wanted something Internet, or software oriented. I tried to tell them, Well there's software in this—it's got a very clever chip in it," Green laughs. "Venture capitalists all follow each other. They always want to spread the risk, they want the other VCs to be enthusiastic as well, so they either all go, or none of them go. That tends to lead to this flocking mentality: Well, if he thinks it's good, I think it's good, too, and they all go charging off in one direction. There's certainly money around, but if you're off on the wrong tack, fighting the wind, it's hard."

Initially, Green raised money by making presentations. Some angel investors, including one of his former partners, chipped in. "It was tough that first stage, but once we got rolling as a public company"—Carmanah was listed on the Vancouver Stock Exchange in 2001—"we found it quite straightforward to raise money."

✳   ✳   ✳

Navigation and aviation lights may just be niche markets, but Carmanah's core idea of coupling LEDs with solar cells also has plenty of other outdoor applications beyond buoys and runways. With the availability of high-power white LEDs, one example is bus shelters. The impetus for this application came from Ken Livingstone, the controversial mayor of London, himself an indefatigable traveler by public transport, who pledged to improve the service and safety of buses in the English capital. LEDs are used in bus stops and shelters in several ways. They illuminate the timetable, to let passengers know when the next bus is due. They also provide enough light to read by at night. Finally, they can signal the driver of an approaching bus that there is someone waiting to be picked up. Through the use of clever power-management techniques, Carmanah's solar shelters are designed to provide light even in the depths of London's notoriously dark and dreary winter.

The cost savings in terms of power of using LEDs in place of conventional lighting is around fifty dollars for one fixture per year. But the main source of revenue for bus shelters is actually the advertising panels they carry. Again, in order to be seen at night, these need to be lit. To fulfill this requirement, Carmanah acquired a small company based in (relatively) nearby Calgary, which specializes in edge-lit LED signage. In September 2005, after a design and development process lasting four years, Carmanah was awarded a contract to supply London with a minimum of fifteen hundred solar-powered, LED-illuminated bus stops. Other cities in the United Kingdom are planning to follow suit, as are, at least to some extent, cities in North America. However, as Green laments, "We're way behind—here in BC, we don't even have schedules on bus stops. The level we're at is, just to put a light there so that people feel safe enough to use the system at night."

Having acquired the ability to light advertisements, Carmanah has moved on to illuminate street names and highway signage. For such applications, fluorescent lamps are conventionally used. These of course have to be replaced on a regular basis. Thus the logic of using LEDs here is the same as for traffic signals, a large and rapidly growing (100 percent per year in 2002) market that Green somehow managed to overlook. "I curse myself. I was focusing on navigation lighting and I didn't spot the traffic lights application. It just didn't occur to me."

Possible compensations loom. For example, down the road apiece comes an even larger application: LED street lighting. Here, however, Carmanah will face stiff competition from much larger firms. For example, in Europe in 2005, Philips installed LED street lights in the Dutch town of Ede. The lamps are expected to last for fifty thousand hours, that is, eleven-plus hours a day for twelve years, four times longer than normal street lighting. Admittedly, these lights are not solar powered. But in Japan, Sharp, the world's largest manufacturer of solar cells, has commercialized stand-alone solar-powered LED street lights that require no overhead nor buried power cables, so they can be installed and turned on in a single day. They cost $4,500 each, which sounds like a lot, but it's much cheaper than digging up the sidewalk.

*   *   *

Dave Green, George Mueller, and Neal Hunter were all fresh out of college when they started their first companies. By contrast, when Don Evans started Cyberlux, he had basically retired and moved to the resort town of Pinehurst, North Carolina, intending to play golf. An economist by background, Evans had participated in starting up several successful ventures, mostly in medical electronics. Then, in 1998, it came to his attention that there was this gap in the marketplace. Evans just couldn't help himself; his entrepreneurial juices started to flow.

"What happened was, there was a study done by this agency," Evans explained in his pronounced North Carolina drawl. "We're in an area along the Atlantic seaboard all the way down the Florida peninsula, then west into Texas and northeastern Mexico, that has a recurring storm phenomenon every year, from June the first to November the thirtieth, called the hurricane season. What is uniform throughout this horrendous experience is power outages. They had studied these power outages, and there was an incidence of, on average, 3.8 days without power that was directly attributable to these storms. The study asked individuals and business owners how they were affected by the loss of power in their homes and businesses. It wasn't in the script, but 80 percent of the people said, Why can't somebody come up with a light that will last for at least five to seven days on one set of batteries?

"Well, I happened to know the people that ran the survey, and they called me because they knew I was always interested in electronics. I went up to North Carolina State University, which is known for its very talented engineering staff, and I sat down and told the good professor there that I was looking for a light that was capable of illuminating a space twelve by twelve feet. I wasn't interested in a focused beam. I wanted to broadcast a blanket of light, but I wanted to do that for a period of forty-two hours on one battery set. He said, How did you arrive at forty-two hours? I told him, Well, the hurricane season occurs in the middle of daylight saving time, it really doesn't get dark until after seven o'clock, so if you had six hours of light from seven o'clock at night, that means you would be able to produce a good room-filling light for a period of six hours a night, for seven days, a full week. So he got on his blackboard, which ringed the room, and he started off here. He went all the way over there, he got to down the end, and he said, It just can't be done."

But Evans is not the sort of person to take no for an answer. He called an old friend from his navy days, an MIT graduate in electrical engineering who had made a pile of money in fiber optics. "He told me, You need to look into the new white LEDs, that is wherein your solution lies. Resultantly, I went to work trying to identify engineering firms that were familiar with white LEDs. There weren't any. Then I found an industrial designer in Florida who had worked for BorgWarner, designing instrument panels for boats that he backlit with LEDs. But they were amber and red, that type of thing. Between the two of us, we got our hands on the Nichia white LEDs. They really provided the solution. And that was in 1999."

Cyberlux, the company that Evans founded in May 2000, has not looked back since. Its first product was a multipurpose LED-based emergency light called EverOn. Designed to be hung from a picture hook on a wall, it contains six white diodes and four amber ones. They run on four AA batteries. How long the lamp lasts depends on how it is used. It has three settings: as a bright room-filling light, it lasts thirty hours; as a reading light, sixty hours; and as a night light, providing a diffuse but reassuring glow, a whopping five hundred hours. The company supplied fifty of these lights to relief workers clearing up after Hurricane Katrina in Biloxi, Mississippi. The workers were asked to return the lights, but none of them did. Evans was disposed to see this philosophically, as a pretty strong endorsement of the product. The EverOn is now available in retail stores. It reportedly sells particularly well during the hurricane season.

Cyberlux has subsequently branched out from emergency lighting into other market segments. They include LED accent lighting strips for under-cabinet applications in kitchens and bathrooms, as a replacement for compact fluorescents; also, for display cases in jewelry stores, as a replacement for halogens. (Retailers are sick of their staff burning themselves on hot halogen lights.) The company is based in Research Triangle Park, conveniently just down the road from Cree.

Obtaining funding for the start-up was, as always, a struggle. In the immediate aftermath of the dot-com bubble bursting, venture capital was well-nigh impossible to find. So, like the other LED entrepreneurs we have encountered, Evans initially approached angel investors. In August 2003 the company went public. Since then, Cyberlux has been able to use its stock to raise money. Around the time of the IPO, Evans brought in Mark Schmidt to take over day-to-day running of the firm. As a fifteen-year IBM veteran, it was natural that Schmidt should see similarities between the rapid evolution of the solid-state lighting industry and that of the personal computer. But there was also a big difference. "This time, we don't have to educate people on the value proposition around light, or energy efficiency, or not having to worry about maintenance," Schmidt explained. "Everyone has that perspective. Everyone innately understands the difference between light and dark."

*   *   *

Carmanah and Cyberlux are just two examples from the vanguard of a seemingly ever-swelling army of solid-state lighting start-ups. Others include BridgeLux, Dialight, Emissive Energy, Gallium Lighting, iLight Technologies, io Lighting, Lamina Ceramics, Lednium, Ledon Lighting, Lexedis, Lighthouse Technologies, Lighting Science Group, Light-Wedge, Luminus Devices, NuaLight, Optek, OptoLum, Phoseon Technology, SemiLEDs, Super Vision, and, as we shall see in part 4, Permlight. In addition, there are also dozens of Asian LED start-ups, most of them based in Taiwan, Hong Kong, and, recently, in mainland China. Almost all aim at providing lighting for the advanced markets of the developing and the developed world. In a sense, however, the most important application for solid-state lighting is in underdeveloped countries, to bring about a radical change in the lives of the poorest of the poor. One man has done more than anyone to demonstrate how to bring about this change. The next chapter tells his story.

CHAPTER 9

Dave Irvine-Halliday is a man who, given half a chance, likes to run up mountains. He used to climb them, too, but a fractured leg in an ice fall and two broken neck vertebrae in an avalanche effectively put a stop to his career as a serious mountaineer. Despite these mishaps, Dave retains his love of mountains; for him, it's a spiritual thing. Not surprising then, that in 1997, when fate presented him with the opportunity to hike the Annapurna Circuit, a 185-mile Himalayan trail that winds around some of the most magnificent mountain scenery in the world, Dave grabbed it with both hands.

A Scottish-born professor of electrical engineering at the University of Calgary in Canada, Dave had been on sabbatical in Nepal doing a project sponsored by the World Bank. For three and a half weeks he had been working with his counterparts at Tribhuvan University in Kathmandu, helping them set up an electrical engineering degree course. It occurred to Irvine-Halliday that while he was there, he could give a one-day course on his specialty, fiber optics. His suggestion was enthusiastically received. It would mean staying a little longer, but he had an open ticket, so there would be no problem. Or so he thought; changing flights from Nepal back to Canada turned out to be much more complicated than he had imagined. Dave found himself with some unanticipated extra time in the kingdom. "I thought, OK," he said, rubbing his hands together gleefully at the recollection, "now I can have a look at Nepal!"

Having considered, and rejected, a trip to see Mount Everest, Dave opted instead for the Annapurna Circuit. This was attractive because, being a circuit, it would mean not having to retrace his steps. He hired a local, Babu Ram Rimal, to act as his guide and porter, and off they went. One afternoon, with the light beginning to fade, the two arrived at a village. Rimal invited Dave to have a cup of tea at a house belonging to two of his relatives. Inside the modest mud-and-stone dwelling, it was very dark. Since the power grid does not extend to this remote part of Nepal, there was no electric light. Dave had always prided himself on the acuity of his night vision. Even on the darkest nights he could usually see enough not to fall over things. Here, however, as he sipped his tea and chatted away, with Rimal acting as interpreter, he was simply not able see his hosts. Except for every now and then, when the wood fire they were using to boil water would flare up, and he would get a flash of light. It was only later, when taking their leave outside, that Irvine-Halliday was able to discern what the elderly couple looked like.

A few days later, in another village near the end of the trail, Dave had a second epiphany. He heard some children singing. Setting off to look for the source of the sound, he came across a school. Although it was daytime, there were no kids around. He peered through a tiny window. Inside, the classroom was very dark. How could the kids see to read? he wondered. Then immediately afterward, for some reason, another thought crossed his mind: I'm a photonics engineer—is there anything I can do to help? It was, as he would often remark to his wife, Jenny, the kind of thought that, once you've had it, you can never forget. During the remaining couple of days of the trek, and on the flight home, Irvine-Halliday pondered what he could do to help the poor benighted people of Nepal.

Back in Calgary, over the next couple of years, at nights and on weekends, Dave gnawed away at the problem. Understanding from the outset that the light source would have to be low power, he tried to produce white light using LEDs. As a specialist in fiber optics, he knew a great deal about semiconductor light sources. But his expertise was with the infrared LEDs used in communications, it did not extend to the visible, high-brightness type of devices. At first Dave experimented by combining red-green-blue indicator-type diodes. The results were pathetic: in the darkest room, even allowing half an hour for your eyes to become accustomed to the dark, you could hardly see the faint circle of light the LEDs emitted, never mind read by it.

Then, one day, Dave happened to be messing about doing Web searches on his computer. For what he swears must have been the umpteenth time, he typed in the words "white LED." Seemingly from out of nowhere, up popped the site of a Japanese company called Nichia, on which there was a picture of a white light emitting diode. Excited by this discovery, Irvine-Halliday immediately reached for the phone and called the company's US office. "If your white LEDs are half as good as you're claiming," he told them, "they could be the answer for this project I'm working on for the developing world." Within days, he received a package of samples.

Pausing only to grab a technician named John Shelley, Dave dashed off downstairs to his lab. There, they hooked up one of the diodes to a power supply, placing directly underneath it on the bench a sheet of typed paper. Dave turned out the lights and the pair waited in the pitch darkness for what seemed an eternity for their eyes to get accustomed to the dark. Then he flicked the switch. A soft white glow lit up the entire paper. "Good God, John," Irvine-Halliday gasped, "a child could read by the light of a single diode!" It was a bona fide eureka moment. The white LED consumed less than one-tenth of a watt. "I knew then that I was onto something." Indeed he was. What he was onto was something that promises to have profound implications for the one-third of humanity—two-billion-plus people—that live their lives off the electric grid.

For such people, nightfall means darkness. To light their homes, villagers in remote parts of the world like Nepal typically use kerosene. To obtain it, they walk for hours, sometimes days, to the depot and back, lugging heavy jerrycans. Recurring expenditure on kerosene eats up a significant fraction of the family budget. But as a light source, a kerosene bottle lamp leaves a lot to be desired. The light from the lamps' wick is so poor that children can only see their schoolbooks if they are almost on top of the flame, directly inhaling the toxic smoke, which is a nasty mixture of carbon monoxide and carcinogens. Smoke from kerosene causes respiratory diseases, eye infections, and lung and throat cancer. It is especially bad for children with asthma.

In addition, kerosene bottle lamps are easily upset. The first time Jenny Irvine-Halliday saw one she thought that it looked like a Molotov cocktail, an accident waiting to happen. Bottle lamps are often blown over by the wind or knocked over by small animals. The fires that result cause terrible burns. Once engulfed in flames, clothing soaked in kerosene can take ten to fifteen minutes to extinguish. The victims are mainly women and children. A mother trying to save her child often gets burned as well. Victims typically receive burns to large parts of their bodies. Approximately 40 percent of burns occurring in Sri Lanka, for example, are caused by the accidental breakage of kerosene bottle lamps. According to a World Health Organization report, burns account for a third of hospital admissions worldwide. They cause 3.5 million deaths annually.

*   *   *

In May 1999 Dave returned to Nepal together with Jenny and some prototype light fixtures that he had built using white LEDs that he had bought from Nichia. Other components, such as switches and batteries, he had also had to buy with his own money. The lights were mostly powered by a pedal-driven generator he designed. The trip was paid for with Jenny's credit card, on the understanding that she got to go with him. They spent three months bouncing around from village to village. Initially, Dave was not sure what the reaction would be. "You'd better prepare yourself for the possibility that people might not like this," he warned Jenny. In fact, the response was overwhelmingly positive. No matter how much Irvine-Halliday tried to talk his LED lights down, everywhere they went, people thought that the lamps were the best thing. Especially since they were so rugged and long-lasting.

One night was particularly memorable. "We were staying in what I jokingly call a minus-seven star hotel. It was this wee guesthouse in a village high up in the mountains. At night, lying on the floor in your sleeping bag, there were these creatures running over you. I had no idea what they were and I didn't want to know. The owner had this pressurized kerosene lantern. It roared like hell, you had almost to shout to make yourself heard above the noise, but it gave a lot of light, far more than the little thing that I had made. I asked him, Can I set up my light? It was like a strip light with eight or so white LEDs, connected to a fully charged battery. I strung it up over a rafter with a piece of string so that it hung just above the table where we were sitting, switched it on, and asked the owner to switch his lamp off. Of course the difference in light was significant. My LED light was a lot less bright, but as we got used to it, the guy realized that there was no smoke, no smell, and no noise. It was quiet. You could actually see quite well, we could all see ourselves, and people were saying, This is great. So at the end of the night, the guy said, Could you please leave this light here? It's much better than what we've got, and kerosene is so expensive."

During this trip, Irvine-Halliday made another significant discovery. "People were using these three D-cell flashlights, but typically the only light their bulbs put out was a weak incandescent glow. Although they couldn't see anything with them, at night they'd walk down these trails with precipitous drops on either side. They have to replace the batteries every couple of weeks, and the bulbs they buy are of such poor quality that they burn out all the time. So I'm looking at these flashlights, and I thought: We can do better than this. I went back to Kathmandu, bought a flashlight, broke it apart, and replaced the bulb with four LEDs, so it was like a wee searchlight. I took it back to the villages, I'd bang it on the table, and the lens would fall out, but the light never went out. I'd hand the flashlight to people, then just as they were about to take it, I'd drop it, and they'd go, Sorry, sorry, sorry. I'd say, That's OK, see—it still works! And they'd want to borrow it to go round the village with, to show it off. Well, the number of times that flashlight disappeared! In the morning, I would say to the headman of the village, I'm not leaving till I've got my flashlight back, and he'd go out and find whoever who had it.

"That's where I came up with the term 'useful light.' What I mean by that is, light that can be used fruitfully by someone in the developing world. Our tests indicate that for a flashlight with an incandescent bulb and three new alkaline batteries, after thirty hours, it's pathetic, but it still glows so we count it; after fifty hours, it's dead. Replace the bulb with three LEDs and at five hundred hours you still have a significant amount of light. At *a thousand hours*, you can still actually read by it. It doesn't light up a whole page, it just lights up a bit. A thousand hours, at three hours a night, is a year. So one set of batteries could actually last you a year, if you really stretch it out. And that just blows you away."

With a single 1-watt LED lamp, it would be possible to light up the most important part of a home in the developing world. Such a lamp would illuminate an area equivalent to a table large enough for four children to sit around and read and write comfortably. Dave returned to Calgary with no further doubts about what he had to do. It took a while, but what had begun in his mind as vague ideas gradually took shape as a well-defined vision. In 1999 he coined a name for his initiative. The title he chose said it all: Light Up the World.*

Almost from day one, Dave realized that if the developing world was going to be lit up, if the initiative was going to go beyond what some people called "backpack philanthropy," then there was no way he could do it working on his own. People in the developing world would have to make it happen for themselves. The technology was in some ways the easy part; in addition, you also needed to have various forms of logistical support in place. There would have to be some "social entrepreneurship," the initiative would be market driven. Dave's son Gregor convinced him that people only truly appreciate something when they pay for it themselves. "The great thing is, there's lots of money to be made from these lights—you don't just give them away."

Dave himself would not make any money. His policy is not to patent any of the technology that he and his co-workers have developed "because, generally speaking, patents keep improvements away from poor people." He decided that wherever it was possible the lights would be produced, assembled, marketed, and sold by local enterprises in the

---

*Light Up the World officially became a foundation in 2002.

country they were used. In Nepal in 2000, for example, with the help of a local entrepreneur called Muni Raj Upadhyaya, who had experience with solar panels, Dave set up PicoPower, a model company located on the outskirts of Kathmandu. In addition to doing assembly work, PicoPower has also exceeded expectations by innovating its own products. "They're on the ground, they do the work, they know what the local people need," said Ken Robertson, Light Up the World's first full-time employee, who joined the fledgling organization in 2001.

Light Up the World's main role is to design white LED lighting systems and to source components—diodes, solar panels, and batteries—testing them to make sure they operate as advertised, then shipping them to where they are needed. Attempting to ensure that the foundation does not pay too much for its parts, Dave came up with the concept of "social pricing." "I told our suppliers, What we want is an agreement where we can buy parts from you at the million-unit price. We become a customer and, regardless of the ups and downs of the market, we can come to you and buy ten thousand or however many LEDs and get them at the million-unit price. Because we're a humanitarian organization, we're not going to be selling them on to somebody who's trying to make a profit."

Remarkably enough, most of his suppliers, no doubt swayed by Irvine-Halliday's formidable powers of persuasion and his record of success in the field, have signed up for the idea. They include LumiLEDs, Kyocera, TIM Industries (a Hong Kong battery maker), and Premier Solar (a manufacturer of solar panels based in Hyderabad, India). The effect of social pricing has been dramatic. "Our batteries went from $20 each down to $4.50," he exults. "I mean, it just turns the economics of the 'light-in-a-box' package as we call it on its head. It makes it so much more affordable for those at the base of the pyramid."

A typical Light Up the World system consists of two or three 1-watt lamps, a 5-watt solar panel, a 12-volt maintenance-free battery, electrical wiring, and some switches. The installed cost of the system averages around one hundred dollars, the hardware accounting for around sixty-five dollars, with shipping, training, customs duties, social impact studies, and other expenses making up the rest. That is without any economies of scale, particularly of the lamps, which include LEDs, lens, circuit board, housing, and the solar panel, the single most expensive part of the package. Considerable scope thus remains for further reductions in price.

But one hundred dollars represents a significant expense for people who only earn a few hundred dollars a year. How can poor people afford LED lighting systems? "With micro-credit," Irvine-Halliday explained. That is, small loans extended to very poor people that are typically used to generate extra income.

There was more to having a reliable source of light than met the eye, though. "What it took this thick skull a year or two to really appreciate was that, this is a double whammy. Once these folk have paid for their LED lighting, they've got something that's much healthier, much safer. It gives out better light, lasts much longer, it's portable, it's rugged, all these kinds of things. So their kids can see to do their homework, which raises the levels of literacy and education, and women can run little cottage industries, which leads to economic and social progress. But wait a minute: What are they going to do with the sixty bucks they used to spend on kerosene? Because every year thereafter, they've got this extra cash. I'm not a sociologist, but I started asking people, What do you think the socioeconomic ramifications of this are? And the answer is, they're immense.

"With the right people in the right place at the right time, you can get exponentials on exponentials in terms of growth. In twenty years, maybe even ten, the developing world could light itself up. Because the prices are coming down and we've got the economic model to get it there. But, having said that, we need help, and we need a hell of a lot of it." In mid-2006, Light Up the World had just two full-time employees (not including the founder, whose only income is his professor's salary) plus two part-timers, and perhaps half a dozen volunteers. Dave Green of Carmanah was a proud supporter of the foundation. "He contacted me, said he really loved what we were doing, and asked if there was any way he could help." The upshot was that Carmanah is paying for a grad student to assist Light Up the World in its work.

In 2005 Dave was invited down to the University of California at Santa Barbara to give a talk. His host, the Nobel laureate Walter Kohn,

Top: Light emitting diodes. (Courtesy of Light Up the World)

Bottom: Shuji Nakamura *(left)* and Nobuo Ogawa *(right)* at Nichia, May 1995. (Courtesy of Bob Johnstone)

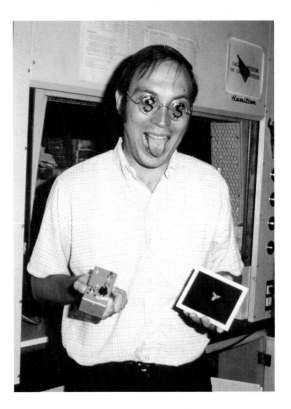

Top: Herb Maruska at Stanford University in 1972, wondering whether blue LEDs will ever make him rich.
(Courtesy of Herb Maruska)

Left: Jacques Pankove at the University of Colorado at Boulder, September 1994.
(Courtesy of Bob Johnstone)

Below: Hiroshi Amano *(left)* and Isamu Akasaki *(right)* at Meijo University, May 1995.
(Courtesy of Bob Johnstone)

Left: John Edmond at Cree, October 2005.
(Courtesy of Bob Johnstone)

Bottom: Shuji Nakamura at the University of California at Santa Barbara, October 2005.
(Courtesy of Bob Johnstone)

Top: Dave Irvine-Halliday with LED lamps in Sri Lanka. (Courtesy of Light Up the World)

Left: Kerosene bottle lamps. "They look like Molotov cocktails." (Courtesy of Light Up the World)

Top right: George Mueller at Color Kinetics in Boston, March 2005. (Courtesy of Bob Johnstone)

Bottom right: Takarazuka School of Art, Osaka, Japan. Lighting design by James Turrell; LED lighting system by Color Kinetics. (Courtesy of Nacasa and Partners)

Lexel-driven Tempura spotlight. (Courtesy of Zumtobel)

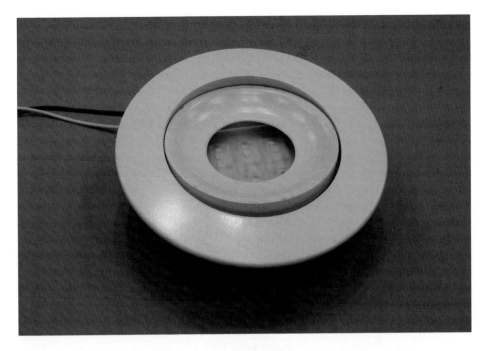

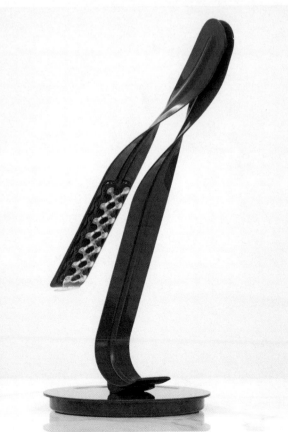

Top: LED can trim. (Courtesy of Permlight)

Left: Leaf lamp. (Courtesy of Herman Miller)

Full Moon Tower, Galaxy Park, Tianjin, China. (Courtesy of TIR Systems)

asked him if he knew where Shuji Nakamura worked. "I think he's in the States somewhere," Dave replied uncertainly. Kohn smiled and said, "Would you and Jenny like to have lunch with him?" Forty-five minutes later, at the UC Santa Barbara faculty club, an astounded Irvine-Halliday found himself being introduced to Shuji. "So, Light Up the World—what is that?" Nakamura wanted to know. Dave embarked on an explanation. Jenny was sitting opposite Shuji. As she told Dave afterward, she noted the expression on Nakamura's face gradually changing, from indifference to passionate interest in what Dave was telling him. Shuji was deeply impressed to have met someone who was actually carrying out his dream, of bringing LED light to the people of the world. For his part, Dave was thrilled to have met the man without whom there would have been no Light Up the World. Before long, the two had struck up a mutual appreciation society. They parted with Nakamura promising to help Irvine-Halliday in any way he could.*

* * *

The University of Calgary is located on the northwestern edge of the city, which some call "the Houston of the North" because it is the hub of Canada's oil industry. From the campus, you can see the snowcapped Canadian Rockies, about an hour's drive to the west. I already knew Dave Irvine-Halliday was a fellow Scot. Where I grew up, a double-barreled name was a sure sign of an upper-class pedigree, so I was not quite sure what to expect. On arriving in Calgary I called Dave from my hotel to fix a time for our meeting. I was surprised and delighted to hear a working-class accent that I instantly recognized. Irvine-Halliday was born in 1942 in Perth, but raised in the Scottish east coast port of Dundee, literally just down the road from my own birthplace. His father, Jim, was a machinist; his mother, Frances, a mill worker. The hyphenated name comes from the fact that, when Dave and Jenny got married, instead of her changing her name to his, they both adopted each other's surnames, an egalitarian practice much more common now

---

*Nakamura was as good as his word. See postscript on p. 301

than it was then. A former midwife, Jenny remains Dave's imperturbable constant companion, always on hand to bring him back to earth when his flights of fancy get too wild.

In person, his attire—or rather, the lack of it—suggests that Dave has a metabolism that burns faster than that of ordinary mortals. On a chilly morning in early April he wears a short-sleeved polo shirt, short pants—revealing heavily scarred knees—and sandals with no socks. He sports this outfit year-round, even during the winter rigors of minus-35-degree Alberta. This human dynamo's standard mode of locomotion is to run everywhere. He is a compact, wiry man with a full beard and a brush-cut head of salt-and-pepper hair. His manner is warm and friendly. He has a ready laugh and a roundabout manner of speaking, sprinkling his conversation with amusing anecdotes told at his own expense.

Dave is above all a humble person who has stumbled on a grand challenge that he is pursuing, passionately, to the best of his not inconsiderable ability. As befits someone who is thoroughly enjoying his life, "living his dream," as he puts it, an inner happiness seems to shine from within him. Which is not to say that everything always goes smoothly for Dave. His dual schedule as professor and founder is punishing, and he is perhaps not the best organized of men. He has a tendency to disconcert his staff on occasion by disappearing off into the developing world, where he remains incommunicado for weeks at a time.

To drum up funding for Light Up the World, Irvine-Halliday has given hundreds of talks at schools, universities, conferences, and organizations like Rotary Clubs. Initially, donations came in dribs and drabs, twenty-five dollars here, fifty dollars there. As the foundation started to make its presence felt, winning prestigious international awards for its work, checks for ten or twenty thousand dollars started to appear in the mailbox. But the goal is grand and the money never goes far enough.

*   *   *

In 2000 Light Up the World lit up its first four villages in Nepal, around 135 homes in total. The following year, the foundation broadened its activities to include India and the island nation of Sri Lanka, at the opposite end of the subcontinent. There, together with a team of dedicated undergraduate volunteers from a local university, Light Up the World has been working in a national nature reserve known as the Knuckles Range.

"It's a very environmentally sensitive area," explains Ganesh Doluweera, a Sri Lankan graduate student in Dave's lab who has helped coordinate the project. "There are some villages deep inside the reserve which will never get any sort of electric utility because you're not allowed to put the grid into such a biodiversity hotspot. If you look at what the villagers do for a living, it's mainly farming. Their kids have to walk a few miles every day to go to school. It could be three or four hours' walk on average, which means the kids have to get up at five o'clock in the morning and they don't get back until four in the afternoon. Within no time it will be dusk, and in order to read or to study, they have to use some sort of artificial light source, mainly kerosene. Although the price of kerosene is subsidized by the government, it still costs a lot, and the price keeps going up—an average person will spend about three dollars a month on kerosene out of a monthly income of thirty dollars. It's a real burden."

Ganesh and his team of volunteers installed their first LED lighting systems in 2003, in about a hundred houses in two villages. Each house was given two lamps and a battery, with each village receiving two battery-charging stations. Initial results were encouraging. "Especially the women. They keep on saying that the financial saving means a lot to them, because now they can buy more food or clothing, or maybe get something for their kids, even get their houses renovated. Some of these people had never seen a battery or any form of electrical lighting before. But they quickly got used to it. The women have to work in the fields on their crops during the daytime, but because they have lighting at night, they might make mats from reeds. They can work on things like that, which they didn't want to do with kerosene because it's expensive. Now they save the money they used to spend on kerosene, they can do some work and earn more money, so their income has almost doubled. They also like taking the batteries to the communal charging stations to be recharged, because it gives them a chance to meet their friends and have a gossip!"

*   *   *

The Knuckles Range project grew slowly to encompass around six hundred homes in twelve villages. Then, suddenly, an unexpected opportunity to implement solid-state lighting systems on a large scale in Sri Lanka arose, in the shape of the tragic aftermath of the December 26, 2004, tsunami. This natural disaster killed hundreds of thousands, and left hundreds of thousands more homeless. Irvine-Halliday spent the days after the tsunami tirelessly working the phone, calling his most generous donors and close friends, asking them for help. Within a very short time he had raised almost $140,000, enough for two thousand LED lighting systems.

As in Nepal, Dave insisted on using local talent. He recruited a small firm, Crystal Electronics, based in the Sri Lankan capital of Colombo, as a partner to assemble and test the systems. "All the lamps are being made in Sri Lanka, which represents a huge influx of capital to that society; it's creating local wealth, which is what you want. We're bringing in solar panels and batteries at a social price from India and Hong Kong. Nichia has donated about thirty thousand diodes and we've bought another thirty-five thousand ourselves. Although these systems are going into refugee camps, they're portable, which means that when the refugees go to their new villages, they'll take the systems with them. It'll be just like we'd lit up the village like we do in our normal projects." At the time of this writing, around fifteen hundred lighting systems had been installed in camps all along the devastated coast, with the other five hundred assembled and ready to be deployed.

In addition to Nepal, India, and Sri Lanka, Light Up the World and its partners have provided light to about fifteen thousand homes worldwide, impacting the lives of more than one hundred thousand people in twenty-six countries. These include Afghanistan, Pakistan, the Philippines, the Dominican Republic, Mexico, Guatemala, Ecuador, Bolivia, Peru, Ghana, South Africa, and Zambia. To spread the word still further, Dave has worked up graduate and undergraduate courses in renewable energy-based solid-state lighting for students in disciplines ranging from engineering to development studies. Thus far, he has been teaching the courses himself with the help of graduate students like Ganesh. He has also been handing out DVDs containing all the course materials "so that they can be used in schools and universities, and by NGOs [nongovernmental organizations] and governments, to help people appreciate what solid-state lighting is truly all about."

The course does not overemphasize the technology. "You've got to understand Ohm's law, but that's about it," Dave says. It touches on such topics as sociology (the economic and cultural benefits of solid-state lighting) and carbon credits (earning income for replacing kerosene). "We compare LEDs to traditional forms of lighting—incandescent, fluorescent, and so on—describing the pros and cons of solid-state lighting. We're quite frank about the fact that it's not all positive. There are some negatives, but it's mainly positive, by a long way. As far as I'm concerned, solid-state is the lighting of choice for the third millennium; LEDs will take over. I don't even discuss it anymore. You no longer say if, it's just when."

CHAPTER 10

In a plenary talk he gave at a conference in Strasbourg, France, in June 1998, Shuji Nakamura made what would turn out to be an overly bold assertion. The essence of what he said was, LED research is over: there is nothing left to do. In the audience that day was Nakamura's good friend, Asif Khan. Afterward, Khan went up to him and said, "You know, Shuji, that's a great challenge—I'll show you that LED research is not finished, that there are still things to do."

Back at his base at the University of South Carolina, Khan considered what actually remained to be done in nitride-based LEDs. In addition to picking up the gauntlet that Nakamura had thrown down, Khan was also motivated by a second consideration. Namely, that as a professor leading a research group, he could not merely be repeating what others had done. He always had to be thinking what the next frontier would be. In a sense, the answer was obvious. The history of light emitting devices has been, as we have seen, a progressive shortening of the wavelength—from infrared, to red, to yellow, to green, to blue and violet—and a corresponding increase in the bandgap, from narrow to wide. That left ultraviolet, "black light" as it is sometimes called, because UV is invisible to the human eye. (But, as any nightclub goer knows, ultraviolet light causes white clothes and teeth to glow in the dark.)

Ultraviolet light subdivides into several categories. There is UV-A, wavelength 380–315 nanometers, also known as long wave; UV-B, 315–280 nanometers, aka medium wave; and UV-C, below 280 nanometers, aka short wave. Big LED companies like Nichia and Toyoda Gosei had focused on making gallium nitride devices that emitted wavelengths at around 380 nm. These LEDs could pump a phosphor that converted ultraviolet to white light. They could even, as Toyoda Gosei had quietly shown, be used in conjunction with a photocatalyst to produce air purifiers as an option for Japanese luxury cars. But thus far no one had ventured below 380 nm, into deep UV territory.

The reason was simple. To get to shorter wavelengths meant moving to a different alloy, aluminum gallium nitride. And, as Khan knew well from his experience, AlGaN was extremely difficult to grow. Add aluminum to the mix and it was a different story, complete with its own distinctive set of technological barriers. Aluminum nitride itself is an insulator, not a promising starter material for devices that rely on the conduction of electric current. In addition to which, the more aluminum you add, the more cracks in your material you get. Not good because, to widen the bandgap sufficiently to reach deep UV, the light emitting layers of devices would need to contain a great deal of aluminum. On the other hand, if you could actually make aluminum gallium nitride LEDs, then a whole new world of applications would open up. In particular, deep UV LEDs had the potential to replace the conventional source of ultraviolet light, mercury lamps. Semiconductor light emitters would, as usual, be smaller, would draw less power, and last much longer. They would also, in this case, be environmentally friendly, unlike mercury, which is a potent neurotoxin (not a nice thing to have leaching from landfills into the water supply). Even at very low levels, mercury can damage the central nervous system, the liver, and the kidneys. Mercury is especially hazardous to pregnant women and young children.

Asif Khan was, he felt, intuitively good at figuring out areas that would be fruitful to pursue. He had after all picked up gallium nitride back in the early 1980s when no one else in the United States was working on it. He and his group had been the first to demonstrate that it was possible to make quantum wells out of the nitrides. But he had been way ahead of his time. Through a combination of bad timing, bad management, and bad luck, in the race to make visible gallium nitride light emitters, Khan had lost out to Nakamura. Now, his instinct told him that deep ultraviolet LEDs were going to be big. "I'm going to go with my gut feeling that this will be an important area," he told his students. "Sometimes you have to take a gamble, you know?"

Khan would bet on deep UV LEDs. In this research, ironically, he would wind up competing with Nakamura and his group at University of California at Santa Barbara. But this time, Khan would win.

*   *   *

In some ways, the achievements of Khan's group at the University of South Carolina mirror those of Nakamura at Nichia. Here, too, modification of the way gases are introduced into the MOCVD reactor jar was crucial. Modification avoided the formation of film-destroying chemical "snow," aluminum being especially eager to react with ammonia. Also vital was the ability to come up with recipes for complex device structures, for depositing stacks of quantum wells that would shepherd recalcitrant electrons and holes to their light-producing destinations. Happily, the group included Wenhong Sun, a world-class crystal grower with a Nakamura-like ability to bend MOCVD systems to his will.

After plugging away at the problem for three years, in mid-2001 Khan's group finally made a breakthrough—an LED that produced appreciable amounts of ultraviolet light at 340 nanometers. They had demonstrated that UV LEDs were feasible. Having made something that worked at UV-A wavelengths, over the next few years the USC researchers pushed on, modifying their approach, retuning their reactors to develop devices that emitted UV-B and UV-C light at wavelengths all the way down to 250 nanometers.

The breakthrough occurred at a propitious moment. Shortly afterward Khan happened to be visiting the headquarters of the Defense Advanced Projects Research Agency at the Pentagon. In a corridor—the Pentagon is mostly corridor—he happened to bump into John Carrano, one of the agency's bright young program managers, whom he had previously met at a nitride workshop. Carrano asked Khan what he was up to. Khan told him about his exciting UV LED results. "John said, Perfect timing, because I'm about to start a large program on deep UV light emitters for biological agent detection, and I would love you guys to participate."

DARPA's Semiconductor Ultraviolet Optical Sources (SUVOS) program would run for five years, with funding of around $50 million. Seventeen projects would be funded, with Khan's group at South Carolina, Nakamura's group at Santa Barbara, and Cree among the principal recipients of Department of Defense largesse. For this program, Khan hired Herb Maruska, who had grown the first GaN films at RCA Labs back in the 1960s. Maruska designed and built a hybrid reactor that the group at USC would use to grow thick films of AlN and AlGaN to provide substrates for ultraviolet LEDs.

The military had three reasons to be interested in black light. One is that deep ultraviolet light is germicidal. It kills viruses and bacteria like *E. coli* (and, were there not a protective ozone layer to absorb it, ultraviolet radiation from the sun would kill us, too). This means that UV light can be used to sterilize water and make it drinkable, a handy trick for troops out in the field. It is not practical to carry mercury lamps, which are bulky and fragile, into battle. But a swizzle stick tipped with a solar-powered UV LED would fit into any soldier's backpack. A company called Hydro-Photon based in Blue Hill, Maine, is already working on a commercial version of such a portable sterilization device.

Covert communications were a second area of interest. Since essentially no deep UV light from the sun reaches the earth, "solar-blind" wavelengths provide an excellent, interference-free window for certain applications. These include networks of ground-based sensors that can detect activity—like bad guys moving around exclusion zones—even in broad daylight.

But the most important reason for wanting a UV LED from DARPA's point of view was as a sensor capable of detecting biological warfare agents, notably anthrax. This was an application that acquired a sudden urgency in September 2001, with the arrival of letters containing anthrax powder at the offices of several US senators and media outlets. Five people died as a result of exposure to this noxious stuff.

To determine what a substance is, whether harmless particle or deadly poison, you shine ultraviolet light on the sample and measure the resulting fluorescence. Conventionally, this is done using extremely expensive, laboratory-based gas-laser systems. With the advent of deep UV LEDs, it became possible to build a portable biosensor system about the size of a laptop PC, costing a fraction of the price. No wonder, then, that Carrano would describe the results of his SUVOS program as "wildly successful."

* * *

"It won't take you long to find the university," Warren Weeks had told me, and it didn't. On a sultry October morning I drove into Columbia, South Carolina's pint-sized capital, passing through rundown neighborhoods of two-story, pastel-colored clapboard houses where, it seems, only African-Americans live. A few blocks away from the city center, I arrived at a massive, apparently windowless, triangular concrete pile. It houses the University of South Carolina's Department of Electrical Engineering, which Asif Khan chairs. The man himself turns out to be fifty-five years old, of slight build and medium height, with salt-and-pepper hair and large, hooded brown eyes. His manner is no-nonsense, but friendly and engagingly frank.

Asif Khan was born in Meerut, the former British army cantonment north of New Delhi, where the Indian Mutiny of 1857 began. His father's people were mostly either teachers or doctors; his maternal grandfather ran a business that made brass musical instruments such as flutes and tubas. Khan's father was a statistician at the Indian national weather service. He was posted to Lahore in what, after the Partition of India in 1947, would become Pakistan. Although most of Asif's relatives remained in India, the name Khan is actually Pathan in origin. This suggests that the family must at some stage have migrated down from the northwestern frontier region of Afghanistan.

On graduating from the University of Karachi with a master's degree in nuclear physics, Khan came to the United States, to MIT, where in 1979 he got his PhD. He joined the Honeywell Research Center in Minneapolis, which in those days was one of the top laboratories working on compound semiconductors. As it happened, Honeywell has a big business making industrial furnaces. The pilot flame in such furnaces emits UV light. A UV sensor, with its narrow window of solar-blind sensitivity, can clearly detect whether the pilot flame remains lit.

Honeywell was keen to replace the conventional flame detectors, photomultiplier tubes. These are expensive, fragile, short lived, and require a high voltage to operate. The company threw the problem to its research center. Gallium nitride UV detectors seemed like a possible solution. So, by a twist of fate, when Khan decided to work on UV LEDs at USC, which he joined in 1997, he was returning to the theme that had been his original point of entry into professional research.

Today, Khan heads an organization that is in the forefront of nitride research worldwide. It is an impressive setup, as I discover when Asif takes me across the road for a tour of his pride and joy: the converted three-story warehouse in which are located the laboratories that he founded and directs. As we move from one room packed with high-tech gear to the next, I ask him whether it is difficult to operate down here in Columbia, a long way from the nearest high-tech nexus. "On the contrary, it's an advantage," Khan tells me. For example, he is not bothered by red tape regarding regulatory issues such as environmental concerns that would be the case were his labs located in the likes of Silicon Valley. Asif has garnered much support from South Carolina's powers-that-be. The local authorities are evidently delighted to have lured a researcher of his caliber to their state. They see him, rightly, as an engenderer, from whose center of excellence good things—prestige, companies, jobs—can be expected to flow.

As we walk from one building to the next, I press Khan on the cultural, as opposed to operational, privations of life in the Palmetto State. Does he miss cricket, a sport to which almost everyone from the subcontinent is passionately devoted? "Not at all," Asif replies. Though he himself has not played much recently, the game flourishes in the southeastern United States. There is a local league, with each city fielding its own team, traveling from place to place each weekend to play. Ah, the unexpected fruits of globalization!

My tour of the labs ends, leaving me with the feeling that I have seen up-close a great thrumming engine of research and development in action. Driven by a dynamic individual, the engine is powered by a phalanx of applied brainpower, measured in professors (five), grad students (twenty), and postdocs (fifteen), all relentlessly pushing the edges of the technological envelope. Many of the staff are not American-born; in particular, there is a large contingent from Asia. I cannot help wondering how many of them will eventually return to Taiwan, Korea, and China to guide the solid-state revolution in lighting as it takes off there.

✳ ✳ ✳

One useful metric for the value of the outcomes of a laboratory is whether any start-ups have formed to commercialize its research. As Bob Davis's lab at North Carolina State begat Cree, so it seems Asif Khan's laboratory at South Carolina begat Sensor Electronic Technology. In fact, the story is more complicated than that. SET was actually founded in 1999 in upstate New York, by two professors from Rensselaer Polytechnic Institute, Remis Gaska and Michael Shur. Gaska had previously worked for Khan as a research scientist at APA Optics, an ill-starred start-up that specialized in nitride technology. The original idea was that SET would develop high-performance, nitride-based electronic devices such as transistors and sensors. Khan signed on as a consultant to the fledgling firm.

When they heard that Khan had decided to go after UV LEDs, Gaska and Shur were initially dubious about his chances of success. But Khan kept badgering them about it, showing them his group's results, urging them to switch their focus to UV LEDs, insisting that there was tremendous commercial potential in such devices. Gaska and Shur remained skeptical: they knew the formidable problems involved in fabricating aluminum gallium nitride devices. But Khan's demonstration that UV devices were doable finally convinced them to get on board.

In 2002, having secured funding from DARPA, Gaska and Shur moved all their operations from Troy, New York, to Columbia, South

Carolina, in order to have close cooperation with Khan's group. The relationship between start-up and university is close. SET takes advantage of the university's expensive tools to analyze its materials. Khan and his longtime collaborator Jinwei Yang are both part owners and directors of the company. Five of its twenty-one employees are former students and postdocs from USC, including Jianping Zhang, a leading crystal grower.

SET took the USC technology and ran with it, working on the performance of the LEDs, improving their efficiency. In May 2004 the company's researchers made a major change in the device design that enabled them to increase the output power of deep UV LEDs by as much as twenty times. This brought the light up to commercially significant levels. They filed patent applications. By the following August, SET had research-grade sample products ready for market.

* * *

It takes about ten minutes to drive from the university, through more dilapidated streets, to the outskirts of Columbia where, off an unkempt side road, the facilities of SET are located. They consist of two aluminum-siding warehouses; more like big sheds, really. An unlikely setting for a high-tech outfit, perhaps, but no more so than that of Nichia, in remote Anan. Outside the company's front door is parked an army green Hummer. The state motto on the behemoth's license plate reads "Smiling Faces, Beautiful Places." Not a bad vehicle to drive, I reflect, if a large chunk of your funding comes from the US Department of Defense.

The Hummer belongs to Remis Gaska, SET's president and CEO. A tall, hefty, blond fortyish man, he is resigned to people asking him about the origins of his unusual name. Remis is in fact a shortened version of *remigijus*, which apparently means "rower" in Latin. Born in Lithuania, Gaska moved to the United States in 1992, after the collapse of the Soviet system. He has two PhDs: one, in physics, from Vilnius University; the other, in electrical engineering, from Wayne State University in downtown Detroit.

People from the Baltic states tend to be laconic—life is hard up there—and Gaska is no exception. But there is no mistaking the excitement in his voice as he explains the extraordinary developments that are taking place in UV LEDs. "It's a revolution," he says. "You'll see!"

*　*　*

Ralph Waldo Emerson famously remarked, "If a man can . . . make a better mousetrap . . . though he builds his house in the woods the world will make a beaten path to his door." He might well have been talking about SET and its range of UV LEDs. As the world's sole supplier of sub-350-nanometer devices, without any advertising or marketing, merely by word of mouth, SET has already attracted more than three hundred customers. Every week, several more beat a path through the woods of South Carolina to the company's door.

Ultraviolet LEDs turn out to have an astonishing variety of applications, including some the company never initially imagined. One difference between the markets for visible and invisible LEDs is that, whereas in the former the exact wavelength of the light does not matter so much— if it's blue, it's blue—in the latter, customers often have very specific requirements. SET is focusing on about a dozen different wavelengths, from 365 nm down to 250 nm. Gaska walks me through the main applications for the devices.

One range is between 365 and 340 nanometers. There, a lot of the interest comes from the curing—rapid drying and hardening—of polymers. These could be inks, or fiber optic coatings, or dental fillings. The small size of UV LEDs is also of interest to the semiconductor industry, to cure the photoresists used to create circuit patterns in microchip lithography. These are all multibillion-dollar markets.

Another range is between 340 and 310 nanometers. This is a prime area of interest for medical applications. Since skin is particularly sensitive to 312–310 nm, dermatologists use UV light to treat skin conditions such as psoriasis and certain allergies. In addition to these existing markets there are also emerging medical markets, such as photodynamic drug therapy, which is used to treat cancer. Once injected, such drugs concentrate in the cancer cells, where they are activated by shining a light on the affected area. Here again the small size of UV LEDs is a major advantage, since it means that biomedical equipment can be miniaturized and made portable for on-the-spot therapy.

A lot of interest in the range between 300 and 295 nanometers comes from the biological research community. As we have seen, bioagents such as proteins are intrinsically fluorescent when excited with UV light. Current lamp- or laser-based optical spectroscopy systems are typically cumbersome and very expensive. A UV-LED-based system could reduce the cost by an order of magnitude, from several hundred thousand dollars to several ten thousands. Such systems would find uses in forensic science and the drug development industry.

Finally, there are also plenty of applications for deep UV light devices from around 280 to 254 nanometers. This is the solar-blind region, where light from the sun does not penetrate. Radiation below 290 nm destroys DNA. "So now you have a tool that is good for disinfection, decontamination, and sterilization of air and water with all the advantages of being a digital source," Gaska explains. What are these advantages? One is that, because a mercury lamp requires time to warm up, it must always be switched on—even when you are not using it. An LED, by contrast, produces instant light because it turns on in less than a nanosecond. Having to keep mercury lamps always on means you have to replace them more often, because they burn out. LEDs are intrinsically long-life devices to begin with. They don't have to be kept on all the time, so they effectively last forever. In military jargon, LEDs are "fire and forget."

Disinfecting water by destroying microbes with ultraviolet light has been used for many years, primarily at bottling plants. More recently, the technology has been adopted by some municipal water districts. LEDs would offer significant advantages for this application, such as dramatically lower running costs (electricity being the largest). However, it will be probably be some years before UV LEDs can match mercury lamps for the sheer output power needed to purify large amounts of water. Here, however, is an example of where the technology can radically alter the nature of the application. Why disinfect at the reservoir? Would it not be better to disinfect at the point of use, in the home? After all, no self-respecting terrorist is going to waste his time chucking anthrax into the municipal water system, where he knows it will get killed. It makes more sense to attack an unprotected part of the system, such as the water pipes leading from the supply. But stick a UV LED in a home faucet and it will zap anything nasty that comes in.

"You can have distributed networks," Gaska predicts, "where you do your personalized water purification whenever you need it. Say you have a coffee maker. You'll have three UV LEDs installed there and they will do the job. You'll get your gallon of sterilized water in twenty minutes. Instead of going and buying gallons of water from the grocery store, you'll have a homebuilt system that you'll know for sure kills bugs. And this is a big problem, because with filtration you don't get rid of bugs, you remove particles and stuff, but not viruses or bacteria. Instead of a fifty-dollar mercury lamp, you can put in a tiny two-dollar LED, and you don't need much power. Once we get to the point when we can make these LEDs much cheaper and more efficient, we keep increasing the flow. Then, as the technology matures, we will go from a quarter gallon per minute, to half a gallon, to a gallon." And on and on, to bigger and bigger applications. One company, Ohio-based Oh Technology, is already using SET's LEDs to treat raw sewage on an experimental basis. Early tests have proved viability, reducing bacterial levels by 60 percent.

"The applications are there," Gaska insists. "For example, look at what's going on with the air and water supplies in airplanes. When you have 350 people packed into one place, and one person is sick, you're not putting mercury lamps up there, that's for sure. But LEDs? No hazard source, no high voltage, no mercury, no toxicity. And you're not talking about one or two planes—there's thousands of them. So we clearly see that UV LEDs are going to be really big. From the application standpoint, it will enable a lot of applications, and a lot of markets. At this point we are creating those markets, because we are providing people with new technology, and they're going through their design cycles. Now customers are coming back, and orders are gradually increasing. Already we are receiving orders for hundreds of thousands of devices. Once volume starts picking up, there are no fundamental reasons why the devices should be expensive. I think UV LEDs will be comparable to visible LEDs. We've showed everybody that it's feasible." Other companies are now following the trail that SET has blazed. "We're a moving target," Gaska concluded, in anticipation of the competition. "We will not be alone."

Part 4 discusses the ongoing revolution in solid-state lighting more broadly. First, however, we must return to witness the dramatic second act in the life of the man who launched that revolution, Shuji Nakamura.

Part Three

FLIGHT OF THE GOLDEN GOOSE

CHAPTER 11

Following Shuji Nakamura's unprecedented series of breakthroughs in bright blue light emitting diodes and lasers, glittering prizes came pouring in from all over the world. Shuji won prestigious awards from Japanese, American, and European magazines, foundations, scientific organizations, and professional societies. But this international recognition from his peers and well-wishers was not matched by raises and promotions from his company.

As we have seen, Nakamura had devoted himself for many years to his research, taking virtually no time off. He had developed revolutionary products. These had doubled Nichia's sales and quadrupled its profits. In addition, on the company's behalf, he had filed applications for hundreds of patents, some fifty of which would prove to be commercially significant. He had repaid his debt to Nichia many times over. Despite all this, Shuji was still a mere section chief earning more or less the same as any other salaryman of his age.

An American company would have most likely made Nakamura a fellow. As such, he would have been honored by his workmates and had the freedom to pick and choose the research themes on which he wanted to work. To reflect a grateful firm's appreciation for his contributions to the corporate bottom line, his paycheck would have been boosted, and he would have been granted a stack of stock options. But not in egalitarian Japan, and certainly not at stingy Nichia.

His new American friends were the first to alert Shuji to this discrepancy between achievement and reward. Sociable by nature, Shuji found it easy to make friends at the international conferences he now regularly attended. After the presentations, he would chat with his acquaintances. "You must be earning a fortune," the more inquisitive among them would suggest. Shuji would smile sheepishly and reply, "Guess how much." His American friends would venture some seven-figure number in US dollars. Then Nakamura would tell them the miserable truth. And they would stare at him dumbfounded. "Why do you stay at such a company?" they would ask incredulously. "If you were working for a US company, you would have been a millionaire long ago." In recognition of his willingness to work so hard for so little, the Americans gave Shuji a nickname.

They started calling him "Slave."

But though his friends made fun of him, and though he was increasingly dissatisfied with his lack of recognition at Nichia, Shuji could not quite bring himself to quit the company where, by 1999, he had worked for twenty years. That year he turned forty-five, said by some to be the prime age for an entrepreneur. Where would he go? With his juniors Nakamura often discussed the possibility of leaving Nichia to form their own firm. When it came to the question of obtaining funding, however, the discussion would founder. Japan has little venture capital available for start-ups. So, though unhappy with his lot, Shuji kept his head down and got on with his research.

Having succeeded in developing bright blue and green LEDs, he was reassigned to work on blue lasers. Like the loyal employee he was, Shuji did as he was told. In the course of his laser research, he managed to reduce the thinness of the light emitting layer by two orders of magnitude, from 1,000 angstroms to just 10 angstroms. Such ultrathin layers would also increase the amount of light output by LEDs. Nakamura went to the LED development department to tell them about his discovery. His former juniors there gave him the cold shoulder. Shuji was mortified. Nonetheless, acting out of a sense of duty to the company, he applied for

As it happened, their middle daughter, Fumie, was already in the United States, at college. Hitomi, the eldest, then twenty-one, was living away from home, studying at a university in Osaka that specialized in foreign languages. Arisa, the youngest at fourteen, was still in high school. A tomboy, she loved sports, basketball in particular. Nakamura was most concerned about Arisa, because she would have to transfer to a US high school.

He told them about the job offer. "Let's go," they chorused. "It's too good a chance to miss." When Fumie came home from the United States for the summer holidays, she added her voice to the chorus. Every day, all three would beg him to let them go. In the end, the female members of his family pushed Shuji to take the plunge. He made up his mind: they would make a new start in the United States.

That fall he started asking the American academics he knew whether UCLA was a good place to for him to go. Naturally, they told him, No, come to our place instead. Soon the word was out—Shuji Nakamura wants to leave Nichia! It spread like wildfire. Almost immediately the job offers started pouring in. Ultimately, ten US and two European universities would attempt to hire Shuji. A scant ten years after he had been looked down upon as a lowly technician at the University of Florida, now some of the brightest names in the academic firmament, like Stanford and Princeton, were falling over themselves to recruit him. Not a single Japanese university showed the slightest interest. It would be a race to see who would be the quickest to make Shuji an offer he could not refuse.

Having ascertained from his friends that UCLA was not the best place for him, his next question was obviously, Where would be good then? Specifically, which universities excelled in nitrides research? Without exception everyone replied, the University of California at Santa Barbara. It's a good place, nice climate, and quite safe—Americans actually go to Santa Barbara to retire. The only problem was that UCSB had yet to make Shuji an offer.

*   *   *

Santa Barbara—the self-styled "city of red tiles" (they looked orange to me)—is located on a strip of coast about two hours' drive north of Los

a patent. He would look back on this as an example of his self-sacrificing behavior, and his appalling stupidity.

When I met Shuji at Nichia in June 1999, he told me that he felt that his work on light emitters was mostly finished. The big breakthroughs had been made. That April Nichia had begun shipping samples of a blue-violet laser diode for use in reading high-definition DVDs. He was working on more powerful lasers that could be used to write discs also, and in other applications, such as laser printers.

I asked him whether he saw himself as an engineer or a scientist. "Pretty much as an engineer," he replied. "From when I was a kid till the time I entered university I wanted to be a scientist. For a while after I joined this company I aimed at being a scientist, but there was no spare time to read books or papers and, if you don't make things, then you can't build a business, so you have to be an engineer. [But] even now I still want to be a scientist." His wish would soon be fulfilled.

By 1999 Nichia's chairman, Nobuo Ogawa, was effectively out of the picture. Bedridden and almost blind, he was no longer able to come to the company. While the old man was still around, Shuji had always managed to get away with his insubordination. The absence of his protector (after a long illness, Nobuo Ogawa died in 2002, at age ninety) left Nakamura exposed to the whims of the company's president, Nobuo's son-in-law, Eiji Ogawa. As we have seen, Nakamura had disobeyed Eiji's orders to stop work on a blue LED. Since succeeding in that endeavor, Shuji had become an international celebrity. Most of the credit for Nichia's rising sales and profits was attributed to him. It would not be surprising if this had made Eiji jealous. He was after all the company's chief executive. So long as Nobuo was around, Eiji had had to bite his tongue. He could not say anything to Nakamura's face. Now, with the old man finally gone, it would be a different story.

In spring 1999 Nichia established a center for research on nitrides. The stated purpose of the center was to develop non-light emitting devices such as high-frequency gallium nitride transistors. Nakamura strongly opposed this idea, arguing that researchers like his friend Umesh Mishra at the University of California at Santa Barbara already had several years' head start in this field. As a latecomer Nichia would have no way to catch

Flight of the Golden Goose

up, something that Shuji knew all too well from his past experience. But Eiji paid no attention to his employee's protests. Instead, he unexpectedly appointed Nakamura as manager of the new center.

Shuji soon discovered that no one else had been assigned to the nitride research center: he was its only staff member. Then he realized what had happened. He was being sidelined, pushed out of the way so that he could not infect younger employees with his disobedient attitude. The Japanese have term for this cruel practice of giving someone a job that exists in name only. They call the unfortunates *madogiwa-zoku*, "the ones who sit next to the window," gazing outside with no work to do until, eventually, they get the message and retire.

Shuji received confirmation of his fate when he bumped into the president, who whispered in his ear, "You'll have to do everything all by yourself again, won't you, Nakamura?" By late 1999 Shuji was twiddling his thumbs. "My job was reduced to checking and stamping documents [i.e., with his signature seal, to indicate that he had read them]. It was all that I had to do. I felt that I would go senile if I carried on like that. I wanted to resign."

Why would a company choose to marginalize its star researcher, rather than bask in his luster? Why follow a policy that seemed expressly designed to render infertile the goose that had laid the golden eggs? Though the low-hanging fruit in light emitting devices had mostly been picked, there were still plenty of projects that Shuji could profitably have pursued. Unless you factor in personal animosity, the decision to treat him in this astonishingly ungrateful way makes no sense.

Nakamura might have become persona non grata at Nichia. Happily for Shuji, however, there was no shortage of other people eager to hire him. University of California at Los Angeles was first in line, and chomping at the bit.

*   *   *

The initial feeler had come from an acquaintance of Shuji's, a professor at Kyoto University, at a conference in January 1999. "UCLA is looking for a good professor. They want to beef up their materials science department

by bringing in new blood, would you be interested?" Shuji had answe noncommittally. Shortly afterward he received an e-mail from King-N Tu, a Taiwanese-born professor at UCLA, inviting him to drop by the versity for a chat next time he was in the United States. Having mor less finished work on the blue laser, Shuji was once again making round of international conferences. On his way home from one of thes early March, he visited Los Angeles. Tu showed Shuji around and off him a job. Nakamura said he'd think about it. At this stage, the ide leaving his company and his native land to become a professor ir United States did not seem real to him. It would be a huge leap.

Over the next few months, Tu peppered Nakamura with e-mail fob him off, Shuji kept adding to his list of conditions. Never ha taught, he was worried about the teaching requirement. Don't w came the reply, you won't have to give any lectures. Then Naka mentioned that he was concerned about the high crime rate ir Angeles. Tu responded that the university would find him a house safest part of the city. Whatever he asked for, UCLA was willing him have it. The package also included a good salary. Eventually mailed him asking if there was anything more he wanted. That put on the spot. He had no reason to reject the university's offer. He his midforties: if he missed this opportunity, he might not get ar offer. Still, he hesitated.

In the midst of his mental anguish about what course to take, went so far as to do something that was, for a Japanese man, unusual: he asked his wife and daughters what they thought. It wa haps the first time he had ever asked for his family's opinior serious matter.

Hiroko had worked at the Tokushima University kindergarter marrying Shuji. In recent years, she had developed a bad back, ar pational hazard for kindergarten teachers, who have to carry littl dren around. Hiroko had begun to complain that she could not co such physical work. At the same time, she was aware that all was n at Nichia. She often criticized the shabby way the company trea husband. "I don't mind going to the US," she told Shuji, "if our ters agree."

a patent. He would look back on this as an example of his self-sacrificing behavior, and his appalling stupidity.

When I met Shuji at Nichia in June 1999, he told me that he felt that his work on light emitters was mostly finished. The big breakthroughs had been made. That April Nichia had begun shipping samples of a blue-violet laser diode for use in reading high-definition DVDs. He was working on more powerful lasers that could be used to write discs also, and in other applications, such as laser printers.

I asked him whether he saw himself as an engineer or a scientist. "Pretty much as an engineer," he replied. "From when I was a kid till the time I entered university I wanted to be a scientist. For a while after I joined this company I aimed at being a scientist, but there was no spare time to read books or papers and, if you don't make things, then you can't build a business, so you have to be an engineer. [But] even now I still want to be a scientist." His wish would soon be fulfilled.

By 1999 Nichia's chairman, Nobuo Ogawa, was effectively out of the picture. Bedridden and almost blind, he was no longer able to come to the company. While the old man was still around, Shuji had always managed to get away with his insubordination. The absence of his protector (after a long illness, Nobuo Ogawa died in 2002, at age ninety) left Nakamura exposed to the whims of the company's president, Nobuo's son-in-law, Eiji Ogawa. As we have seen, Nakamura had disobeyed Eiji's orders to stop work on a blue LED. Since succeeding in that endeavor, Shuji had become an international celebrity. Most of the credit for Nichia's rising sales and profits was attributed to him. It would not be surprising if this had made Eiji jealous. He was after all the company's chief executive. So long as Nobuo was around, Eiji had had to bite his tongue. He could not say anything to Nakamura's face. Now, with the old man finally gone, it would be a different story.

In spring 1999 Nichia established a center for research on nitrides. The stated purpose of the center was to develop non-light emitting devices such as high-frequency gallium nitride transistors. Nakamura strongly opposed this idea, arguing that researchers like his friend Umesh Mishra at the University of California at Santa Barbara already had several years' head start in this field. As a latecomer Nichia would have no way to catch up, something that Shuji knew all too well from his past experience. But Eiji paid no attention to his employee's protests. Instead, he unexpectedly appointed Nakamura as manager of the new center.

Shuji soon discovered that no one else had been assigned to the nitride research center: he was its only staff member. Then he realized what had happened. He was being sidelined, pushed out of the way so that he could not infect younger employees with his disobedient attitude. The Japanese have term for this cruel practice of giving someone a job that exists in name only. They call the unfortunates *madogiwa-zoku*, "the ones who sit next to the window," gazing outside with no work to do until, eventually, they get the message and retire.

Shuji received confirmation of his fate when he bumped into the president, who whispered in his ear, "You'll have to do everything all by yourself again, won't you, Nakamura?" By late 1999 Shuji was twiddling his thumbs. "My job was reduced to checking and stamping documents [i.e., with his signature seal, to indicate that he had read them]. It was all that I had to do. I felt that I would go senile if I carried on like that. I wanted to resign."

Why would a company choose to marginalize its star researcher, rather than bask in his luster? Why follow a policy that seemed expressly designed to render infertile the goose that had laid the golden eggs? Though the low-hanging fruit in light emitting devices had mostly been picked, there were still plenty of projects that Shuji could profitably have pursued. Unless you factor in personal animosity, the decision to treat him in this astonishingly ungrateful way makes no sense.

Nakamura might have become persona non grata at Nichia. Happily for Shuji, however, there was no shortage of other people eager to hire him. University of California at Los Angeles was first in line, and chomping at the bit.

✻   ✻   ✻

The initial feeler had come from an acquaintance of Shuji's, a professor at Kyoto University, at a conference in January 1999. "UCLA is looking for a good professor. They want to beef up their materials science department by bringing in new blood, would you be interested?" Shuji had answered noncommittally. Shortly afterward he received an e-mail from King-Ning Tu, a Taiwanese-born professor at UCLA, inviting him to drop by the university for a chat next time he was in the United States. Having more or less finished work on the blue laser, Shuji was once again making the round of international conferences. On his way home from one of these in early March, he visited Los Angeles. Tu showed Shuji around and offered him a job. Nakamura said he'd think about it. At this stage, the idea of leaving his company and his native land to become a professor in the United States did not seem real to him. It would be a huge leap.

Over the next few months, Tu peppered Nakamura with e-mails. To fob him off, Shuji kept adding to his list of conditions. Never having taught, he was worried about the teaching requirement. Don't worry, came the reply, you won't have to give any lectures. Then Nakamura mentioned that he was concerned about the high crime rate in Los Angeles. Tu responded that the university would find him a house in the safest part of the city. Whatever he asked for, UCLA was willing to let him have it. The package also included a good salary. Eventually Tu e-mailed him asking if there was anything more he wanted. That put Shuji on the spot. He had no reason to reject the university's offer. He was in his midforties: if he missed this opportunity, he might not get another offer. Still, he hesitated.

In the midst of his mental anguish about what course to take, Shuji went so far as to do something that was, for a Japanese man, most unusual: he asked his wife and daughters what they thought. It was perhaps the first time he had ever asked for his family's opinion on a serious matter.

Hiroko had worked at the Tokushima University kindergarten since marrying Shuji. In recent years, she had developed a bad back, an occupational hazard for kindergarten teachers, who have to carry little children around. Hiroko had begun to complain that she could not continue such physical work. At the same time, she was aware that all was not well at Nichia. She often criticized the shabby way the company treated her husband. "I don't mind going to the US," she told Shuji, "if our daughters agree."

As it happened, their middle daughter, Fumie, was already in the United States, at college. Hitomi, the eldest, then twenty-one, was living away from home, studying at a university in Osaka that specialized in foreign languages. Arisa, the youngest at fourteen, was still in high school. A tomboy, she loved sports, basketball in particular. Nakamura was most concerned about Arisa, because she would have to transfer to a US high school.

He told them about the job offer. "Let's go," they chorused. "It's too good a chance to miss." When Fumie came home from the United States for the summer holidays, she added her voice to the chorus. Every day, all three would beg him to let them go. In the end, the female members of his family pushed Shuji to take the plunge. He made up his mind: they would make a new start in the United States.

That fall he started asking the American academics he knew whether UCLA was a good place to for him to go. Naturally, they told him, No, come to our place instead. Soon the word was out—Shuji Nakamura wants to leave Nichia! It spread like wildfire. Almost immediately the job offers started pouring in. Ultimately, ten US and two European universities would attempt to hire Shuji. A scant ten years after he had been looked down upon as a lowly technician at the University of Florida, now some of the brightest names in the academic firmament, like Stanford and Princeton, were falling over themselves to recruit him. Not a single Japanese university showed the slightest interest. It would be a race to see who would be the quickest to make Shuji an offer he could not refuse.

Having ascertained from his friends that UCLA was not the best place for him, his next question was obviously, Where would be good then? Specifically, which universities excelled in nitrides research? Without exception everyone replied, the University of California at Santa Barbara. It's a good place, nice climate, and quite safe—Americans actually go to Santa Barbara to retire. The only problem was that UCSB had yet to make Shuji an offer.

*   *   *

Santa Barbara—the self-styled "city of red tiles" (they looked orange to me)—is located on a strip of coast about two hours' drive north of Los

Angeles on Highway 101. Its population of one hundred thousand contains a sprinkling of famous residents from the entertainment industry. They include Kirk Douglas, Kevin Costner, Brad Pitt, John Cleese, Bernie Taupin, David Crosby, Kenny Loggins, and Oprah Winfrey.

In 1975, when Herbert Kroemer was invited to join the electrical engineering faculty at UCSB (which is actually not in Santa Barbara proper but in the contiguous town of Goleta, to the north), he found what he described contemptuously as "a Mickey Mouse place." Like every other university, its solid-state electronics laboratory was doing research on the mainstream material of the semiconductor industry, silicon. Kroemer was not impressed. He told his host that his interest lay in compound semiconductors, a field that he thought would become increasingly important in the future. At that point, only three other US universities had "critical mass"—by which he meant more than two professors—in this new field. There was, Kroemer felt, still room for a fourth. "By going on with mainstream silicon there was no chance of being better than second- or third-rate. I've always felt that it's better to get in on the ground floor in something where you can be first-rate. And if you cannot be first-rate, it's better to be absent, because at least that way you cannot be accused of being second-rate," he laughed gruffly.

By 1999 Kroemer had been instrumental in assembling at UCSB what was widely regarded as the number-one group in the world in the compound semiconductor device field. It included six full professors and thirty-odd graduate students. They liked to work together as a big team, sharing resources, students, and labs (a highly unusual modus operandi in academe, where isolated fiefdoms tend to be the norm). One was Steve DenBaars, who joined the faculty in 1991 from Hewlett-Packard, where he had worked on high-brightness red LEDs.

DenBaars recalled how Nichia's November 1993 announcement of the bright blue LED had galvanized faculty at UCSB. "We had started writing proposals for grants to do gallium nitride as early as 1991." This was far ahead—in fact, too far ahead—of the pack. "Everything we wrote got turned down. People said gallium nitride is a waste of a material; but right after [Shuji's] breakthrough, because we'd already identified what we needed, we were able to move quite quickly. So in 1994 we were up and running almost full-on in gallium nitride."

Bernd Keller, then a postdoc in DenBaars's group, remembered those early days as a time of great excitement. "Steve posted this announcement outside the office, and once this was out, we didn't ask for permission, we just did it [saying], We need to do this, we'll worry about the money later. We instantly launched a reactor design project, and within three months we had a reactor going on trying to follow this path. As a result I think we demonstrated the first double heterostructure, the first [gallium nitride] LED of that type in the US."

"Steve used Scotch tape, coat hangers, and paper clips to string together a starting program that was completely inspired by Shuji's results," DenBaars's colleague Jim Speck agreed. "Steve was the first academic in the US to really see what Shuji had done and to understand it, and he ran with it very aggressively. He was able to build this team here, even before Shuji arrived, into the largest and most cohesive nitride research team outside of industry in the world. He convinced Umesh Mishra and me to start working on gallium nitride, and we started working aggressively. We really went all out on it."

"What caught our eye, what got us on the bandwagon was that blue LED was such an amazing thing," Mishra added. "I clearly remember a discussion that Steve and I had. . . . We said, We've just got to jump, abandon what we're doing, change horses in midstream."

DenBaars, Speck, and Mishra had all known Shuji since the mid-1990s. They met him four or five times a year at international conferences. They were used to him bitching about his boss. DenBaars had spent time in Japan, as a consultant to Stanley Electric. He knew that Japanese salarymen were forever whining about how badly their companies treated them. That was why UCSB was not the first to make Nakamura an offer: they simply did not realize that, this time, he was serious about leaving.

In late September 1999, DenBaars visited Nichia to give a lecture. There, Shuji told his friend that UCLA was trying to hire him. He asked for DenBaars's opinion. DenBaars did not hold back: "I was like, WHAT!? Why go to UCLA when you can come to Santa Barbara and work with your friends? If you come to us, we'll help you set up your lab. It's easier for us to raise money. If you go to UCLA, you'll be the only one there doing gallium nitride."

"Our philosophy was, Hey—if he wants to leave, then he should come here as opposed to going someplace else," Mishra said. "It would be good both for him and for us." Kroemer recalled that it was Mishra who came to his office and told him that there was a possibility of getting Nakamura to come to UCSB. "I said, Oh by golly, let's go after it. I was a very enthusiastic supporter right from day one." Mishra also made an impassioned pitch to his faculty, saying "Look, he's like the Michael Jordan of this field. The team he's on wins, and the team he plays against loses." Of course it was a risk—no one doubted his intellectual ability, but would Shuji fit into an academic setting? It was a risk worth taking.

Nakamura made his first visit to UCSB soon after DenBaars's visit to Nichia. The latter showed him around, introducing him to Kroemer and other faculty members. It is hard to imagine anyone not being impressed by the university's lovely campus. The location is so picturesque. The campus juts out into the Pacific Ocean, with spectacular views across the strait to the nearby Channel Islands in the foreground; behind, the Sierra Madre Mountains recede majestically into the distance in deepening shades of brown.

Stylish modern facilities are interspersed with wide-open spaces. There are tall trees including eucalyptuses and Norfolk Island pines, in which humming birds can be observed feeding. A sense of well-being pervades the campus. The calm is interrupted only if you happen to stray onto the paths that connect the various departments. Packs of students on wide-handlebar bicycles come whizzing around the corners, bearing down suddenly upon the unsuspecting visitor. What initially impressed Nakamura most, however, was not so much the look as the scale and the informality of UCSB. The campus was relatively small, unlike UCLA, which was huge. You could walk into the president's office, without an appointment, for a casual chat.

In November Shuji went back to Santa Barbara on his own to check out the city and its environs. He noted with approval the mild climate and the miles of white sandy beaches where young people gathered to play his favorite game, volleyball. On a more pragmatic note, there seemed to be lots of Japanese restaurants in town. The downside was that house prices were very expensive and the cost of living seemed high. As far as

Nakamura was concerned, however, his biggest worry was not whether he could afford to live in Santa Barbara. Rather, it was the fact that, if he decided on UCSB, he would have to give lectures.

In mid-October, he attended a conference in North Carolina where he met his old friend and rival, John Edmond of Cree. Edmond, as we have seen, had been trying unsuccessfully to hire Nakamura for years. Now, Shuji told him that he was serious about leaving Nichia. Edmond made him a very generous offer: a salary of $500,000 a year, plus stock options potentially worth millions. Shuji was inclined to accept. The remuneration was much better than he could expect at a university. And at Cree, he would not have to teach. With his mind at ease, he could devote all his energies to research. He told DenBaars that he had decided to join the company. His friend took the news on the chin, saying that it couldn't be helped; in Shuji's place, he would do the same thing.

At the end of the month, however, Nakamura happened to run into DenBaars again at yet another conference. DenBaars took Shuji aside and asked him whether he had considered the possibility that, if he moved from Nichia to another company, an archrival, he might be sued by his former employer for leaking trade secrets. Nakamura was taken aback. Not well versed in the ways of the world, the possibility of being taken to court had never occurred to him. DenBaars pressed home his advantage: On the other hand, if you move to a university, he told Shuji, there's no danger of being sued because you would not be involved in product development.

Seeking further advice, Shuji called another friend, Jim Harris, a professor at Stanford. "I think you're really much better off going to a university," Harris told him, "because I'm sure Nichia will sue the living daylights out of any company that you go to." Harris was keen to get Nakamura to come to Stanford. North Carolina State was also in the hunt. So were several other universities and companies. But UCSB beat them all to it.

*   *   *

Knowing that Nakamura was now leaning toward a life in academe, DenBaars was determined that he should come to Santa Barbara. He went to see the UCSB's dean of engineering, Matt Tirrell, and the university's chancellor, Henry Yang. "I told them, This is a once in a lifetime opportunity, let's move very fast."

Moving fast is not an easy thing for a university system to do. Especially not one with a huge bureaucracy, like the University of California. And especially not during the 1990s, when UC was facing severe budget constraints, forcing it to offer early retirement to thousands of its professors. Typically an appointment like the one DenBaars was proposing takes at least a year to work its way through the system. Knowing this, Kroemer was initially pessimistic about the chances of being able to hire Nakamura at short notice. Then he realized that, when push comes to shove, the bureaucrats could get a move on. Though no position existed, one was created for Nakamura. "It was amazing how much support we got locally, at every level, including the chancellor," Kroemer recalled. "And we pulled it off!"

In December DenBaars and Mishra flew to Osaka to meet Shuji. They felt that it would be better to make their offer face-to-face. "It was one of those weekend things where you get on a plane on a Friday for a meeting on a Saturday," DenBaars recalled. It was just before Christmas and snow was starting to fall. Even at that late stage it was not certain that Nakamura would accept their offer. Not knowing what they were competing against made it hard to guess which way Shuji would jump. "He had every option in the world," Mishra said. "We knew that, logically, we stood a chance because we were the top program in the US. Nevertheless, the cachet of Stanford is always hard to battle, and he could still have gone to industry, or to some other place."

They need not have worried. Determined not to carry any baggage with him into his new life by joining a company, Shuji signed on the dotted line. As he put it, "Steve and Umesh came to Japan with the contract from UCSB. . . . They flew business-class, stayed in the best hotel, and came home smiling."

*   *   *

Having made up his mind to quit Nichia, in early November 1999 Nakamura began preparing for his departure. He threw everything away except his research papers. His desk and surroundings were normally messy. Now they were tidy to the point of being unrecognizable. Alerted that something was amiss, his juniors began asking him, Are you going to quit?

Nakamura submitted his letter of resignation to the manager of Nichia's development department on December 27, 1999. It was the day before the last working day of the year at the company. The manager said that he would see what the president said. To Nakamura it did not matter what Eiji Ogawa said. He simply told his boss that he would not be coming to work tomorrow, and went home. That night, he received an e-mail from the company. It said, "Tomorrow being the last day of the year, there is an assembly for all workers. We'd like to you to give an address to the company. Then we'll begin the retirement procedure."

In fact, what happened next day at Nichia was that the company tried to force Nakamura into signing a noncompete agreement. Nichia was willing to pay him severance money, but only on the condition that he signed the agreement. Under its terms, he would not be able to work on gallium nitride or file any patents for three years. Luckily, Shuji had been forewarned by UCSB's lawyers. You are not obliged to sign anything, they advised him. Especially not something that might compromise your freedom as a professor to do research. Any legal matters should be handled via UCSB. So Nakamura told Nichia to contact the university's legal department to send them an English-language version of the agreement. And he refused to sign.

In retaliation, the company withheld his retirement compensation. Before their departure for the United States, Shuji and Hiroko sat down and calculated the total income they had earned over the past few years, he as a corporate researcher, she as a kindergarten teacher. It turned out that they had each made almost the same amount. Nakamura would subsequently grumble that Japan was socialist country where everybody received the same reward, regardless of achievement.

A few days later, Shuji took the ferry from Tokushima to Osaka. His juniors from Nichia took the day off work to come and farewell him. The scene was reminiscent of forty years earlier, when his elementary schoolmates came to the pier at Oku to wave him good-bye. Shuji had always got on well with his juniors. He had looked after them, allowed them freedom to do research; in return, they looked up to him. They parted on good terms, with promises of future get-togethers.

At Osaka Airport, a send-off of a very different kind awaited him. In Japan, it is rare for an employee to quit his company in midcareer, and even rarer for a senior employee to leave Japan for a career overseas. As a result of his achievements, Shuji had become something of a celebrity. Reporters and camera crews from no fewer than five television stations were on hand to capture his departure for the evening news. When he got off the plane at Santa Barbara another Japanese TV crew was there, and yet another would be at UCSB to record his first day at his new job.

On January 10, 2000, UCSB put out a press release. "Inventor of Blue, Green, White LEDs and Blue Laser Leaves Japanese Company for US University," ran the headline. "Shuji Nakamura's Research Likely to Lead to a Whole New Way of Lighting." The release quoted an exultant chancellor Henry Yang: "Nakamura's joining our group is like mounting the jewel in the semiconductor crown at Santa Barbara."

Shuji was gracious in his explanation of why he had chosen to move from industry to academe. "I am very grateful to Nichia for the superb research opportunities they have provided me," the release quoted him as saying. "But I felt that I had come to a crossroads in my life. There is a spectrum of . . . research that ranges . . . from the purely commercial to the purely academic. I decided that an academic context would now be better for my work."

He bought a house in Santa Barbara's upscale Hope Ranch suburb. The price was steep, but the university helped arrange the financing for him. For Hiroko, the attraction was the garden; for Shuji, the fact that UCSB was just a fifteen-minute drive. Nakamura formally began his new life as a university professor on February 19, 2000. That October, Herbert Kroemer won the Nobel Prize for Physics, the same month that Shuji obtained his green card and permanent residence in the United States. In December 2000 Nichia sued Nakamura for leaking trade secrets to Cree.

*　　*　　*

CHAPTER 12

Nichia's chairman, Nobuo Ogawa, was determined not to license to other companies the high-brightness LED technology that Shuji Nakamura had developed. His policy went counter to the semiconductor industry's long-established practice of cross-licensing. The basis for this practice was mutual self-interest. In a field where so many bright minds were at work, no one company could hope to maintain a monopoly on intellectual property. (And regulatory bodies like the US Federal Trade Commission would take a dim view if they tried.) Cross-licensing made sense because it meant that firms did not have to waste vast amounts of time and money in court fighting lengthy legal battles. Even companies such as Texas Instruments, which held basic patents on the microchip itself, was prepared to cross-license those patents—for a price. But although Nichia could undoubtedly have made a mint from license fees, from the start the firm flat out refused even to consider the option.

There were several reasons for this no-license policy. One was altruistic: old man Ogawa had always wanted to provide jobs for the good people of Tokushima, historically one of the poorest parts of Japan. Manufacturing LEDs represented an unprecedented opportunity to boost local employment, with the company increasing its payroll manyfold. (By 2005 Nichia was employing some thirty-five hundred people, up from just seven hundred ten years earlier.) Another was fear: Nichia was a small company. Licensing its technology to much bigger rivals could be very risky. With better access to capital, such competitors would find it easier to make massive investments in production capacity and marketing manpower. Nichia might be overwhelmed. A third reason, which Naka-mura explained to me, was very Japanese: licensing was too easy an option. In the long run, it would make the company soft. Better that they should earn their profit the old-fashioned way, by making and selling products themselves. Such a course would ensure Nichia's long-term prosperity, even if that did mean behaving like a monopolist.

Adopting such a hard-line attitude was unheard of in harmony-loving, consensus-seeking Japan. Nichia's litigious approach to all-comers would ultimately end up pitting the company against its most famous former employee in what may turn out to be one of the most important court cases in Japanese corporate legal history.

*   *   *

Nichia's campaign to protect its precious intellectual property had begun in August 1996. In the Tokyo District Court, the company filed a lawsuit against Toyoda Gosei—Isamu Akasaki's commercial partner—which had recently commenced shipping double-heterostructure gallium nitride blue LEDs. Claiming that the Nagoya-based firm had infringed its LED patents, Nichia demanded that its rival halt shipments and pay damages of approximately $5.5 million. The following year, Toyoda Gosei responded with a counterclaim. In 1998 Nichia launched a legal barrage, consisting of a further five suits. Once again Toyoda Gosei responded in kind. A spokesperson for Nichia described the conflict as "developing into an all-out war."

Sumio Shinagawa, an attorney representing Nichia, saw the battle as one of the most complex patent-related disputes ever conducted in Japan. Such disputes were mostly settled simply by the infringing company agreeing to pay a licensing fee. In this case, however, Nichia had no intention of negotiating an agreement. Since its patents related to pn-junction-type LEDs, whereas Toyoda Gosei's intellectual property was mostly related to obsolete MIS devices, Nichia had the upper hand. The company refused to consider Toyoda Gosei's reconciliatory proposals aimed at setting up a cross-licensing deal.

This first battle was one that Nichia would win. In August 2000 the court ruled that Toyoda Gosei had indeed infringed Nichia's patent. It ordered the Nagoya-based company to stop manufacturing the offending device and to pay Nichia approximately $900,000 in damages. (Toyoda Gosei appealed the judgment, and the case dragged on until a settlement was finally reached in September 2002.)

Nichia's next campaign would be a very different story. The first salvo was fired in December 1999, when the company sued Sumitomo Corporation, the Japanese distributor for Cree, again in the Tokyo District Court. Charging patent infringement, Nichia demanded that Sumitomo stop selling blue LEDs in Japan. Cree intervened on behalf of its distributor, taking over the defense of the case. In April 2000 Nichia brought two additional patent infringement lawsuits against Sumitomo.

Then things got complicated. In May 2000 Cree acquired a company called Nitres. This was a start-up founded in 1996 by Steve DenBaars and Umesh Mishra of the University of California at Santa Barbara to develop high-brightness gallium nitride LEDs. The company's primary focus was ultraviolet light emitters. It employed around twenty-seven people, mostly former students and technicians from UCSB. Its facilities were located at a business park adjacent to the university. Following its acquisition, the company was renamed Cree Lighting. The new subsidiary's charter was to do medium-term—meaning six months to one year—development work for the parent company.

In late September 2000 Cree and North Carolina State University retaliated against Nichia by filing a lawsuit in the Eastern District Court of North Carolina. This suit sought an injunction on sales of Nichia's blue-violet laser in the United States. Then, in early November, Cree dropped a bombshell. The company announced that it was hiring Shuji Nakamura to work as a part-time consultant at Cree Lighting. At the same time, Cree moved to strengthen its ties with UCSB by pledging $1.2 million to fund an endowed chair to be known as the Cree Chair in Solid-State Lighting and Displays. The company also threw in another million dollars to fund research at the university. There were no strings attached to this money: Cree would not receive any rights to technology developed as a result of the donation. The following year, Nakamura was named as the inaugural recipient of the chair.

In the United States, it is of course common for professors to consult for companies. Indeed, some schools—MIT, for example—virtually require that faculty spend one day a week working in industry. This is seen as healthy, a way of avoiding the ivory-tower syndrome, of making sure that research does not drift too far from the requirements of the society that, one way or another, is paying for it. If intellectual property results from this mutually beneficial interaction, then so much the better (as long as the university gets its cut). Professors becoming rich is not seen as an undesirable outcome. In Nakamura's case, it was quite natural that he should consult one day a week for Cree Lighting. After all, the company had been founded by his friends, and it was staffed by their former students. Besides, where else in Santa Barbara was he going to consult?

In Japan, by contrast, the ties between academe and industry are far less intimate. For one thing, there is the issue of turf. Universities are the jealously guarded province of Japan's education ministry, which pays professors' salaries. The bureaucrats see no reason why academics should supplement their incomes by consulting for companies, which come under the jurisdiction of a rival ministry. Accordingly, they spin a web of red tape to discourage cooperation. A second barrier is cultural. There is a strong feeling among the Japanese that academics should not sully their lily-white hands through overly close contact with grubby industry. They should do only academic research.

From an American perspective, Nakamura was doing nothing out of the ordinary. From a Japanese one, it was unexpected that a professor should do work for a company. But the worst thing, from Nichia's point of view, was that their former star employee was apparently now consorting with the enemy, their archrival, Cree, the outfit that only the previous month had filed suit against them.

On December 27, 2000, a year to the day after Nakamura's resignation, Nichia filed a counterclaim against Cree and NCSU in the Eastern District Court of North Carolina, alleging infringement of four US patents on gallium nitride light emitters. In addition, the suit cited Nakamura, accusing him of leaking trade secrets to archrival Cree. In a cruel twist of fate, the very outcome that Shuji had joined the faculty of UCSB in order to avoid had come to pass.

✳   ✳   ✳

In May 2001, the Tokyo District Court dismissed the first of the Nichia suits against Sumitomo Corporation and Cree. This came as no surprise to Shiro Shinagawa, Nichia's patent attorney, because, as he told *Nikkei Business*, "Cree has a trump card named Shuji Nakamura, the inventor of Nichia's blue LED." The structure of Cree's device differed from that of Nichia. (And, as Cree chairman Neal Hunter never tired of telling people, their devices were based on silicon carbide.) But Cree had also found another stick with which to beat Nichia. It turned out that Shuji Nakamura had not been the first to patent a gallium nitride buffer layer, one of the key steps in making bright blue LEDs. In 1991, just one week before Nakamura filed for his patent, Ted Moustakas of Boston University filed a similar application. Cree would subsequently take out an exclusive license on Moustakas's US patent, which was granted in 1997. Based on this patent, in May 2001, Cree counterattacked, filing an infringement suit against Nichia and Nichia America.

The Nakamura patents, which some in Japan had referred to as an "unsinkable aircraft carrier," had been torpedoed. The company's monopoly on bright blue LEDs was taking on water fast. In December 2001 the Tokyo court dismissed Nichia's other two claims against Sumitomo and Cree.

Eventually, in November 2002, Nichia would concede and do what it vowed it would never do. The company entered a comprehensive patent cross-license agreement with Cree (also with Toyoda Gosei and the remaining two of the "Big Five" gallium nitride LED makers, LumiLEDs and the German firm, Osram Opto). The long drawn-out legal battle had been, as Cree's John Edmond put it, with characteristic bluntness, "a pain in the ass. Just a huge waste of time and effort, and funds." At the height of the disputes Cree was spending, it was said, a million dollars a month on intellectual property-related legal issues. "In the end," Edmond said, "we just came together and said, Hey, you do what you do, we do what we do, and we'll just cross-license everything." But that was by no means the end of the legal matters.

*   *   *

In Nichia's case against Nakamura, the company contended that Shuji was still bound by a nondisclosure agreement he signed when he joined Nichia in 1979. The suit asserted that Cree had hired him to gain access to knowledge he had accumulated while working at Nichia. The main evidence for this assertion was some e-mails that Shuji had "secretly" sent to John Edmond in late 1999, when he was considering whether to join Cree. Nichia claimed that in these e-mails Nakamura had promised Edmond that he would show Cree how to get around Nichia's patent wall.

Nakamura responded that all he had done was to voice his concern over whether there would be a legal problem with patents if he joined Cree. There was no question of him divulging Nichia's trade secrets to a rival. Nakamura's US lawyer, William McLean, referred to the case as "a spite suit," an attempt to punish Nakamura for what Nichia saw as his treachery.

US courts, unlike their Japanese counterparts, have a process known as "discovery." It obliges the parties in a legal case to disclose pertinent information. Under this process, all e-mails were revealed. If there had been anything incriminating in them, Nakamura would have lost the case, been fined, perhaps even sent to prison. Instead, he was completely exonerated. In October 2002 Nichia's allegations of trade secret theft were found to be baseless and the suit was dismissed. But the court took almost two years to reach this decision. For Shuji, it was all a massive waste of time involving lengthy depositions, cross-country trips for appearances in court to give evidence, and seemingly endless consultations with lawyers. In addition to which, there were some nasty moments, like when he was accused of perjury, an assertion that was quickly determined to be false.

The allegation that he had leaked trade secrets to Cree enraged Shuji. Despite Nichia's mean-spirited refusal to pay him the retirement money they owed him, Nakamura had been content to let bygones be bygones. But his former employer had come after him aggressively. Shuji decided he would retaliate by fighting fire with fire. On August 23, 2001, he filed a claim in Tokyo District Court asking for compensation of 2 billion yen (around $16.5 million) as his fair share of the $1.4 billion in sales that Nichia had thus far earned based on the technology he had developed.

"I had no intention of filing a suit against Nichia when I quit the company," Shuji told Yoshiko Hara of *Electronic Engineering Times*. "All my hope was to start a new research life from square one in the United States. But Nichia's action [i.e., in bringing the trade secrets suit] was a hindrance to my work. It was nothing other than noise. It made me angry and I decided to file a suit from my side." As one Japanese commentator put it sagely, "One cannot help but think that Nichia [now] lies on a bed of its own making."

For an inventor to file suit for fair reward is not uncommon. But Nakamura's petition also contained a most unusual demand. Namely, that 80 percent of the rights to the key patent he filed while at Nichia should be returned to the inventor, i.e., him. The Japanese patent in question, number 2,628,404—usually abbreviated "404"—covered two-flow MOCVD, the novel crystal growth system that Shuji had used to make all of his blue LED and laser breakthroughs. The petition also pointed out that Nakamura had developed gallium nitride LEDs against the instructions of Nichia's management, which had ordered him to stop the work. He had applied for patents secretly, without Nichia's knowledge. They were thus the result of his initiative rather than the company's requirements.

Nakamura said he was claiming the rights in order to make them available for wider use by other manufacturers. "If Nichia had not monopolized the patents, the blue LED market would have grown ten times larger than it is today," Shuji told Hara. "LEDs are energy- and resource-saving devices, but Nichia has stunted their healthy growth." This was radical stuff.

Fighting the suit on his behalf in the Tokyo court was a maverick lawyer named Hidetoshi Masunaga. His goal was not merely to win his client's case. Rather, Masunaga wanted to force a revision of Japan's ambiguous patent law, to bring it in line with the American model. Article 35 of the US patent law stipulates that employees are entitled to receive "adequate compensation" for their patented research. Japanese researchers typically do not have contracts with detailed provisions for rewards. In the United States, researchers' rights are clearly stated in their employment contracts. The greater the contribution of the employee, the larger his share of the earnings. In theory, at least.

In Japan it is entirely up the company how much an individual inventor should be compensated. For each of the one hundred and ninety-plus Japanese patents he had filed, Shuji had received ten thousand yen (less than one hundred dollars) on application and another ten thousand yen on granting. Dispensing such paltry amounts regardless of the commercial consequences of the innovations they related to was standard practice among Japanese companies.

Much more than money was at stake here. At a time when Japan's economy was in the doldrums, the need to stimulate innovation—hence, product development—was urgent. Nakamura's point was that the nation's prosperity depended in large measure on the talent of its innovative engineers. By failing to reward such people adequately, Japan risked losing its technological edge, as the best and the brightest increasingly left to find jobs overseas. In filing such a suit against his former employer, Nakamura felt he was striking a blow on behalf of all Japanese corporate researchers. His action sent shock waves through Japanese industry, especially among senior management and corporate apparatchiks responsible for handling intellectual property. By daring to challenge the status quo, by questioning the wisdom of absolute loyalty to the company, Nakamura emboldened others to assert their rights as individual inventors.

In response, researchers filed a rash of multimillion-dollar suits against giant corporations such as Hitachi, Toshiba, and Ajinomoto for compensation for inventions relating to optical discs, flash memory, and artificial sweeteners. Needless to say, corporate Japan thought that this kind of thing was outrageous. "Product innovation cannot be created by one person," spluttered Toshiba chairman Taizo Nishimuro. It was the joint effort of the corporation and its employees that made products possible. Ultimately, however, Nakamura's bold action would force many Japanese companies, including industry leaders like NEC and Honda, to rethink their remuneration policies regarding innovation.

In September 2002 the Tokyo court rejected Shuji's attempt to win the rights to his patents, ruling that a tacit contract obtained between the parties under which Nakamura gave ownership of his inventions to Nichia. At the same time, however, the court determined that Nakamura was eligible for monetary compensation based on the amount of profit generated by the intellectual property for which he was responsible. The only question now was, How much?

$$* \quad * \quad *$$

Regardless of their outcome, court cases can be bad publicity for those involved. This is especially true in Japan, where it is felt that if a case goes to court, then both the plaintiff and the defendant are to blame. Participants thus acquire a certain notoriety. This can have unfortunate ramifications. For example, when Shuji moved to the United States, NHK, Japan's national broadcaster, produced a ninety-minute TV program about him going to UCSB and the kind of research he intended to conduct there. The documentary was completed about a week before Nakamura filed his suit against Nichia. Following the announcement that he had taken legal action, the program was hurriedly canceled.

Rather than have their good names dragged through the mud of a public trial, most Japanese would much rather settle matters out of court. While Nichia was suing Nakamura for leaking trade secrets and Nakamura was suing Nichia for adequate compensation, in an attempt to limit the damage, efforts were made to reach a settlement.

At UCSB, Shuji and his colleagues had established the Solid-State Lighting and Display Center. Seven companies, most of them Japanese with Cree being the only US member, contributed funds to the center. In return, the firms could each send a researcher to work at the university and get first crack at licensing intellectual property. Now, as a gesture of goodwill, Nakamura decided to invite Nichia to join the center. He called his former company's headquarters to make the invitation personally. What ensued was a near-farcical failure to communicate.

"First I talked to the manager of Nichia's patent division. He said, I can't make that decision. You'll have to talk directly to the president [i.e., Eiji Ogawa]. So I called the president's office, and a woman answered. She asked me to wait a moment while she asked the president. I waited two or three minutes, then she came back and said, 'The president's not here.' I waited a few minutes, then I called back and the same thing happened. Then I called the patent division manager again, and he gave me the president's home number. I called the number and his wife answered. I asked to speak to her husband. 'Is that Mr. Tanaka?' she asked me, and of course I didn't tell her who I was. She asked me to wait. Then, after two or three minutes, she came back and said, 'My husband's not here.' So after that, I gave up."

Being the high-profile protagonist of a cause célèbre inevitably made Shuji a media star in Japan. Newspapers and magazines solicited his opinions, not just on his area of expertise, but also on general topics. Nakamura was prepared to speak out with characteristic frankness from his vantage point outside Japan about what he felt was wrong with his native land. In a country that is unusually prone to navel-gazing, such outspokenness made his columns and op-ed pieces compelling reading. Shuji the rebel technologist became Shuji the social commentator. He would frequently rail against the shortcomings of the Japanese education system. In particular, the pressures put on young people to cram for university entrance exams. These he mocked as being little more than memory tests, "ultra-quizzes," in his word, that left would-be students feeling listless and worn-out. In his view, the system stymied the growth of the individual. It produced salarymen with a sheepish mentality who lacked confidence in their own abilities and who slavishly obeyed the instructions of their companies. "Japan is a good country to live in for those with no ambition," was his conclusion.

Meantime, in court, Nichia was doing everything in its power to portray Nakamura in a bad light. As evidence against him, company employees introduced expert opinion letters in which they asserted that Nakamura had only been one member of a team, that he had not personally been responsible for much. Mostly, these letters were from employees who had joined Nichia after Nakamura's departure to the United States. But there were also some written by Nakamura's former juniors at the company. It was hard for Shuji to read such words by his most trusted associates, whom he had mentored early in their careers. But he understood that they were only doing so under duress. Through his contacts at the company Nakamura heard that Nichia's president had warned these employees that, if they did not toe the company line, they would be fired. With families to support, and little chance of finding other employment in Tokushima, they were forced to comply.

Hoping to dig up some dirt, Nichia even resorted to hiring private investigators to spy on Nakamura. Warned by UCSB's lawyers that he was under surveillance, Shuji became paranoid, worrying that his phone might be bugged and imagining that someone might be tailing him. He claimed that the investigators secretly recorded and videotaped him at work and even in his private life.

Nichia also attempted to dredge up the by-then-refuted allegations that Nakamura had leaked trade secrets to Cree. However, since this matter was entirely unrelated to the current case, the strategy backfired. The presiding judge, Ryoichi Mimura, was irritated by the mention of irrelevant issues.

On January 30, 2004, Mimura awarded Shuji a total of 20 billion yen ($190 million) in compensation. It was by far the largest award of its kind ever made by a Japanese court. The judge commented that Nakamura deserved such a large sum because "the invention was a rare example of a world-class invention achieved by the inventor's unique ability and unique ideas in a poor research environment at a small company."

The court arrived at the staggering figure of 20 billion yen based first on an estimate of the company's sales of blue LEDs from 1994, when the devices went on sale, to 2010. The total sales figure was assessed at 1.2 trillion yen ($11.4 billion). Second, the court estimated that if Nichia had licensed its patent portfolio to other companies, these rivals would have generated an additional 600 billion yen by the year 2010. Assuming patent royalties of at least 20 percent, the court said that by attempting to retain its monopoly on the bright blue LED market, Nichia had forgone 120 billion yen ($1.14 billion) in royalty payments. The judge also ruled that Nakamura's contribution was "not less than 50 percent," adding that his patents "had made possible the commercialization of [bright] blue LEDs." Nakamura was pleased that his contribution had been recognized. "This ruling will increase the incentive for researchers to invent," he commented, "and companies will profit from it over the long run as well."

Nichia immediately appealed the ruling, arguing that the judgment failed to assess the contributions of the company and its other researchers. Now, as the case made its way to the appeals court, Nichia initiated hostilities on a new front, attempting to besmirch Nakamura's reputation in the court of public opinion. In the summer of 2004, a book was published in Japanese- and English-language editions. The title of the English edition was *Blue Light Emitting Diode: Invented by Nichia Corporation and Its Young Engineers*. The anonymous authors of this woeful attempt at a hatchet job are listed as "Themis Editorial Department," Themis being an obscure Tokyo-based publisher (and, in Greek mythology, the goddess of justice, no doubt an intentional reference).

The preface makes perfectly clear the book's intent and its sponsor. The media were biased, reporting only one side—i.e., Shuji's—of the blue LED story. Now it was time to tell "the truth" about who was really responsible: "Behind [Nichia's] research and development of the blue light emitting diode lies President Eiji Ogawa's bold decision. He went against the opposition of those around him and came up with the necessary funding by even taking out a mortgage on his own house to give the project full support." Under the "strong guidance" of Ogawa, the invention and the success of the blue light emitting diode were "the fruits of the hard-working efforts of Nichia Corporation and its young engineers."

The book is in fact a Stalinesque attempt to airbrush Shuji out of the picture, to take away the credit for his achievements. The authors contend that Nakamura was "involved in the development of the blue LED as a member of the technical staff." This is like saying that Lance Armstrong was involved in the winning of the Tour de France as a member of the US Postal Service cycling team. It ignores the fact that breakthroughs are typically made by individuals, not by consensus.

The book accuses Shuji of being primarily interested in making money. Also, of having been un-Japanese for claiming inventions as his own rather than as the product of group efforts. And, of being a traitor to Japan, by working "for the United States." Though many of the allegations the book makes are demonstrably ludicrous, they have unfortunately gained some credence among people who do not know Nakamura, and hence, have done some damage to his reputation. It is therefore necessary to refute some of the most egregious assertions.

For example, the book claims that the annealing method of making positive-type gallium nitride was invented not by Nakamura but by two of his juniors. Nakamura had taken their research results, passing them off as his own. Given that both of the juniors were recent hires, one of them having joined the company only about six months prior to the breakthrough, this is hard to swallow. According to Nakamura: "I instructed them to do the annealing, and they did it under my guidance. They were just technicians doing what they were told to do. That's all there is to it."

Similarly, the book asserts that Nakamura's contribution to the development of the blue laser was "extremely small," since he was away from the lab most of the time, traveling around Japan and the world, going to conferences. Nakamura's response: "It was all done with me as their supervisor, telling them what to do. It's not true that I did everything myself, but it was under my leadership. With the laser, there was a team of about ten people, but the only one who was there from start to finish was me, and I was giving the instructions."*

To clarify matters, I asked John Edmond of Cree for his assessment. Of course it can be argued that, since Nakamura now has close connections with Cree, Edmond's judgment may be biased. Nonetheless, since his career in many respects parallels that of Nakamura, and since he has known Shuji for many years, it seems to me that Edmond is eminently well qualified to comment on Nakamura's contribution.

"Shuji was key, because he did the initial work. After that, he did go to a lot of conferences, but probably as he was gallivanting around he'd pick something up and bring it back and say, OK, guys, let's try this and

---

* For an example of how Nakamura directed laser development at Nichia remotely, see chapter 5, p. 126.

this and this—let's go make it happen. I guarantee he was involved in the experiments, because I know how I was involved—and I went to a lot of conferences, too. You can't *not* be involved, you've got to be involved. I mean, I think Shuji had the most knowledge of what had to happen next. And I guarantee that he had most of the good ideas. Did he have all the ideas? No. I don't have all the ideas. There are several good people in our group. But I think he had the good leadership, to push it along. And to argue that he had nothing to do with it, or that he was very remote from the process, is just craziness."

Shuji's friend and colleague Steve DenBaars remembered that, when Nichia researchers other than Nakamura were making presentations at conferences, they would always defer to their leader. "You'd ask them questions, and even though their English was good enough, they typically wouldn't be able to answer basic science questions; they would say, Nakamura will know. I mean, I went to all his early conference talks and it was clear that he was the only one who understood the technology. I've known Shuji a long time, we've had detailed technical discussions, and he knew gallium nitride inside out. And he still knows it much better than these guys."

Ultimately, at the news conferences that followed the various judgments in the compensation case, Nichia representatives were repeatedly asked by both Japanese and non-Japanese journalists the following, telling question: If Shuji Nakamura was indeed so unimportant in the development of bright blue, green, and white LEDs and the blue-violet semiconductor laser, then why was this point not made until after he had left Nichia? While Nakamura was still an employee, the company was perfectly happy to let him be seen by the world as the main mover behind its R&D. Even as late as 2001, in its trade secrets suit against Cree, Nichia acknowledged that Nakamura's work had made the company a world leader in gallium nitride-based devices. The fact that they knew that his knowledge was so important was precisely why Nichia was so keen to stop him from, as they saw it, consorting with the enemy.

*   *   *

In December 2004 the Tokyo High Court recommended that Nakamura and Nichia get together and work out an amicable settlement. Nakamura's lawyer urged him to accept the advice, on the grounds that the original decision might be reversed if he refused. On January 11, 2005, a settlement was announced. Nichia would pay their former employee 843 million yen (about $8 million).

Both sides declared victory. Since this was the largest amount ever awarded as compensation to a Japanese corporate researcher, Shuji's lawyer claimed that Shuji had won. The rights of an individual versus a company had been established. A precedent had been set. Though unhappy with the terms of the settlement, Nichia decided to accept it and get on with business. Thereafter, Eiji would tell customers visiting Anan that the problem with Shuji was simply that he had wanted too much money for himself, and that was not in the best interests of the market.

For his part, Nakamura's initial reaction to the judgment was that he was not at all satisfied at the outcome. In fact, he was furious at what he saw as a biased judicial system that had sided with corporate Japan against him as an individual. In seeking to cap his settlement at a much lower amount, the Tokyo High Court had, he felt, acted in Nichia's interest to avoid damaging the company's future growth by requiring it to make a large payout. As for the amount he had been awarded, most of it would go to pay taxes and legal fees. He would be lucky if he had enough left to pay off his mortgage.

Later, however, when he had cooled down, Shuji would come to see the settlement as a win, albeit not on the scale that he had hoped for. The issue of whether the individual or the organization was primarily responsible for inventions was still unresolved. But at least he had struck a blow on behalf of the individual inventor. And he was very happy about that.*

Despite the fact that the legal battle was now over, Nichia would continue to do everything in its power to denigrate Nakamura. For example, when he spoke publicly, Nichia people would be there at the venue, distributing copies of their anti-Shuji tract. And they would badger the organizers of his lectures, insisting on their right of reply.

---

* In March 2006 Nichia announced that it was abandoning its rights to the "404" patent. In a statement, the company claimed that it had ceased using the technology covered by the patent in 1997.

Nonetheless, in Japan, Nakamura has become something of a folk hero. He is recognized there wherever he goes. DenBaars recalled a recent visit to Japan with Shuji and their colleague Jim Speck. "Just walking through Tokyo Station, we got stopped by three different groups of people that wanted to have their picture taken with him. The first was a group of mature women; the second, some older salarymen, who wished him good luck. We were teasing him about how famous he was, and he was saying, No, it's only the older people that recognize me. And maybe one minute after he said that, two young girls stopped him and asked for his picture."

It occurred to DenBaars that "a lot of people see his fight against Nichia for reasonable compensation as a kind of movement for change, for cultural reform on how people are rewarded. Shuji always feels that Japan has to change, that the best Japanese baseball players are leaving to play in the US—like Ichiro Suzuki of the Seattle Mariners—and now the best Japanese scientists are starting to leave, too."

The lawsuits had eaten up enormous amounts of his precious time. At their most taxing, Shuji was having to go back to Japan almost every month. Now, with the legal distractions out of the way, as we shall see in the next chapter, he was finally able to focus on his research—and a new element in his life, his students.

CHAPTER 13

Southern California is well known for its laid-back lifestyle. Californians like to blend work and play. They keep surfboards in their offices, take off in the middle of the day to go running or biking or hang-gliding, stuff like that. A lesser man might have been content to rest on his laurels, relax, and have a little fun.

Not Shuji Nakamura.

"Shuji is a guy with an incredible work ethic," his friend and colleague at UCSB Steve DenBaars told me. "He gets in to work before seven in the morning, every morning. In almost five years, I've only beaten him in once—and I had to get in at 5 AM to do it! He works till seven in the evening, incredibly hard hours and zero procrastination; that's why things move fast when you're around him. And he doesn't like his students to procrastinate—if he says to do an experiment, they'd better be doing it that day or the next day or they're going to be in trouble."

One of Shuji's first graduate students, Rajat Sharma, recalled how one night, the group was breaking in a new reactor. They were using it to grow crystal round the clock, and he was the designated babysitter. "At some point around 5 AM, I locked myself out of the lab. After sitting around for a while I was going to head home. It's 6:45. I'm in the parking lot and Shuji arrives. So I got to borrow his access card. But he didn't seem very surprised that I was there."

Another of Shuji's first students, John Kaeding, remembered seeing somewhere a list of five things to do in order to be a successful graduate student. "One of them was always be in the lab before your thesis adviser and go home after him. I looked at that and laughed, because there is no physical way I could work more hours than Shuji—I'd be in a casket by the time I was thirty!"

*   *   *

Getting his new state-of-the-art crystal growth lab at UCSB up and running took Shuji two years. He had had to start from scratch. From the outset, his idea was to branch out into areas that he had not covered at Nichia. With big companies now driving mainstream nitride research, it made sense to aim for niches. Things like alternate growth techniques, improving the quality of materials, and building far-out device structures. That way it would be possible to do real science and technology. MOCVD, Shuji's pet technique, had become the growth technology of choice for most devices. But there were things that MOCVD was not good at, such as high-pressure growth. Accordingly, Nakamura spread his net widely to encompass several other crystal growth systems, such as molecular beam epitaxy (MBE) and hydride vapor phase epitaxy (HVPE).

The low-hanging fruit had been picked, most of it by him and his group at Nichia. Reaching the high-hanging fruit would take longer and greater efforts. Such seemingly inaccessible objectives included growing bulk gallium nitride, that is, producing ingots of crystalline material. As we have seen, all commercial gallium nitride LEDs and lasers are grown on sapphire or silicon carbide substrates. But the mismatch between materials produces high defect densities and cracking between device layers.

Having substrates of flawless gallium nitride would be wonderful, especially for lasers, which are particularly vulnerable to cracks that drastically reduce their lifetimes. Gallium nitride substrates would make it possible to complete the set of semiconductor lasers, including the missing link, green. Red, green, and blue lasers could then combine to produce a beautiful display, the ultimate color television, with full-spectrum rendering and ultrasharp images. (RGB LEDs are already beginning to replace fluorescent lamps as the backlights in LCD TVs. LEDs are superior because they produce deeper hues across a broader range of color.) But growing ingots of gallium nitride is extremely difficult. Like bulk diamond, the process requires very high pressures. A high-quality two-inch wafer of gallium nitride currently sells for around ten thousand dollars a pop. And even that is not proper bulk material. (It still has to be grown on a substrate made from a different material.)

Another outstanding challenge is to improve the performance of green indium gallium nitride LEDs. These currently have far lower electricity-to-light conversion efficiencies than blue and red devices (less than 10 percent, versus more than 40 percent, respectively). This failing is somewhat mitigated by the human eye's sensitivity to light, which peaks in the green part of the spectrum. But when red, green, and blue devices are combined to produce white light, compensation cannot be by eye. Extra green LEDs must be added (e.g., in a backlight, you need two green LEDs for each red and blue). For all sorts of reasons, this is unsatisfactory.

Building up the infrastructure in which to conduct research took time. First of all, Shuji had to find funding. Much of the money came from Japan. In 2001 Nakamura and DenBaars formed the Solid-State Lighting and Display Center at UCSB. Seven firms, five of them Japanese, plus one Korean and one American (Cree), each chipped in $2.5 million in membership fees. In April 2002 Shuji won a grant from the Japanese government's Exploratory Research for Advanced Technology program to investigate bulk gallium nitride. It was worth $16 million over a period of five years.

*   *   *

In May 2004 I went to see Shuji in his new home. I drove up the coastal strip from Los Angeles to Santa Barbara. Eye-catching mauve jacarandas were then in full bloom, as were Southern California's distinctive coral trees, their spiky, lipstick-red flowers vivid against the backdrop of—appropriately enough under the circumstances—a bright blue sky.

On my way to Shuji's office, early for our appointment, I bumped into the man himself. He was just dashing off to give a lecture (the fourth time around, Shuji told me later, he was finally getting used to lecturing). It had been five years since our last meeting, almost ten since our first. Dressed in an old yellow polo shirt, faded jeans, and scuffed-up black shoes, Shuji looked casually disheveled. His hair was receding and graying now. But the characteristic vivacity I remembered was very much in evidence, and he still punctuated his utterances with that inimitable, infectious high-pitched laugh. How was Shuji enjoying life as a professor? I wondered. "It's good, you're free to do anything you like," he replied. "And there's no boss to tell you what to do, ha-ha-ha!"

At that time, with the compensation suit still under way in Tokyo, Shuji was obviously distracted by legal matters, which we discussed at some length. Eighteen months later, the court case finally over, he seemed much more relaxed, relishing his life in academe. But he was still incredibly busy, popping in and out of meetings with students, faculty, and visiting donors, preparing for a forthcoming trip to China, now in his office, now in his lab, now elsewhere.

In addition to having to lecture, the other big difference between a corporate researcher and an academic scientist is that professors have students to nurture. On my second visit to UCSB, in October 2005, I made a point of spending time with some of Shuji's postgrads, asking them what it was like to work for him. After all, having been in the same lab together with Nakamura day-in, day-out for several years, they knew him better than almost anybody. (Especially since no one who knew him at Nichia would be allowed to speak freely about what Shuji was like during his time there.)

*   *   *

Ben Haskell was one of Nakamura's first grad students at UCSB. "When I interviewed here, it was about a week after Shuji arrived. To be honest,

I was totally ignorant about gallium nitride, and I had no idea who Shuji was. But I was talking with Steve [DenBaars] and Jim [Speck, another professor in the Materials Department], and they said, You've got to talk to this guy. So they brought me into his office, and it was just piled high with boxes. He hadn't even unpacked, only had half his furniture. We just talked for a while, and clearly he was new to the game. I was one of the first grad students he talked to, but even in the first few minutes we talked, he became much more comfortable with me and I became very comfortable with him. And it didn't take a whole lot of coaxing after that to decide to come here and work for the three of them.

"In my first year, I didn't do much because the lab was under construction, but towards the end of that year, Shuji started to guide me towards hydride vapor phase epitaxy of gallium nitride. And again, it was something that I didn't really know anything about, but he steered me in that direction, gave me tremendous freedom to try all sorts of experiments using that growth technique. . . . It was sort of a new area for him, too, so he was learning as we were going along also. But he just had tremendous insight, at every step of the way. So it was really a nice environment to work in, because you would have his knowledge on the one hand, plus his manner, being so easy to get along with, as a mentor and an adviser."

John Kaeding was another grad student who encountered Shuji early on. "I sat down to meet with Shuji, on his second day of work I believe. He gave me a brief intro to his research ideas, areas he was going to work on, with a little overview starting with a schematic representation of the reactor that he designed, and how that enabled him to make these breakthroughs. And then he said, You'll probably need to design your own system. So I knew you weren't going to be walking into a situation where it was, OK, this is the setup, here's the manual, learn how to use it, just turn the crank and you're going to get a result. It was more along the lines of, You're going to have to take responsibility for everything.

"If we have an idea that the equipment we're working with isn't performing the way we want it to, he has no problem with us radically changing this multimillion-dollar machine. He lets us, because we're the ones in the lab, we're the ones using it day-in, day-out. He kind of listens to our intuition, and he lets us do things that aren't successful and that sometimes maybe waste time. But he gives us that freedom to follow in his footsteps, so to speak, attempting to break the mold."

John Kaeding's friend and fellow student Rajat Sharma concurred. "A lot of the ideas he suggests require changes to our hardware, which take a few days to make. The idea in the end may work or not work, but that's never an impediment to our trying it. Instead of spending ten hours debating whether something will work or not, his approach has always been, just try it and you'll know—you have so much time on your hands, you have the reactor twenty-four hours a day, if you think something might work, just try it. Shuji's going to be constantly giving you ideas. And they're not always good ideas, I mean, nobody is so brilliant that everything they say is gold. It's more a problem of too many ideas, and too few people, and not enough equipment or time to work on things."

"Shuji's not a big theory guy," Kaeding said. "All of his intuition is based on what he knows and the science he knows, but at the same time, he gives you all these ideas, and some of them are bad. I think that's because a lot of it is, he thinks about something and he's willing to immediately try and implement it. And once you start peeling away the first couple of layers of the onion, you might realize that this is not really going to work, or not going work the way we thought it would. But he's not the sort of person who's going to sit down for weeks trying to do calculations and play mind games, and say, Oh, this is definitely going to work, let's do it; he likes that more engineering approach."

The transition from corporate group leader to academic thesis adviser has not been without hiccups, however. Paul Fini, the lead researcher in Nakamura's laboratory, told me about Shuji's initial frustrations. "The way I always envisioned it, before Shuji came to UCSB, at Nichia he would have a team of, let's say, twenty or thirty researchers under his control. He would be generating ideas, and they would pick them up and run. He would say, Jump! and they would say, How high? So the cycle of development was very quick, and Nichia has always been known for that.

"When he came here, he was probably thinking that same sort of model would work, which it doesn't. Ben [Haskell] is a good example of that—Ben is the only one on his machine, period. One grad student running one machine, that really limits the amount of results you can kick out in a given amount of time. So the Japanese model had to be altered in Shuji's mind. But I think he adapted pretty quickly to the ways we do things here."

I asked Fini and Haskell whether they considered Shuji a leader. "Absolutely," Fini replied. "He's a leader, and he's a good professor, because a good professor in my mind doesn't have to be in the lab every day, as long as they have oversight of the project, and they're an ideas generator. And Shuji's definitely an ideas generator, in a very good sense. His style is laissez-faire, he puts a bunch of ideas out there, makes sure things are on track, but he doesn't look over your shoulder and say, Shouldn't you be trying this? or, Shouldn't you be doing that?"

Shuji had to learn to be laissez-faire, though. "When he first got the lab going, he wanted to be running the machines himself!" Haskell recalled, laughing. Fini agreed: "I do remember him coming in and talking to John and Rajat multiple times a day. But that was when essentially they had no clue about MOCVD growth. They needed a lot of guidance. He did start out trying to get the ball rolling, but now he's much more relaxed in his management style."

"Shuji used to drive us fairly hard, especially in the early days when we were just getting things going," John Kaeding said. "You'd get five calls on your cell phone, and he'd pop into your lab three times; he'd come into the lab and talk to you for half an hour, walk out, then thirty seconds later call you with a new idea. Although that can sometimes be almost too much, he's not one of those professors where you need to call his secretary and schedule an appointment that's going to be two weeks later. You can walk up to his door and knock on it. I've come in when he's on the phone with someone in Japan, and he quickly hangs up. Sometimes I almost feel bad because it's such a small stupid question that if I'd been willing to take a couple of hours I could have found out the answer on my own. Instead, I just walked over to his office, knocked on the door, and he answered my question for me in five minutes."

Even away from the lab, Shuji is responsive to questions from his students. "He's always been very good about keeping in touch by e-mail," Fini said. "He'll check anywhere in the world where there's a wireless connection. I've got responses from him more quickly when he's away than when he's here. If you do need something critical answered, like, Should we try this, should we try that, what shall we do here? you'll probably get an answer back in six hours. So even when he was traveling a lot, it still worked out."

"Sometimes you read war histories that talk about a general as being a soldier's general," Kaeding said. "In that sense, Shuji is very much in the trenches. Not so much now, but especially in the early days, he'd be in the lab, I've been there with him, both on our hands and knees, trying to diagnose some problem with the machine. And rather than simply talking to us or looking at the data, when you show him a physical sample, you can almost see his eyes light up. He is very active in what goes on. So I can't see him sitting at a desk at Nichia, simply rubber-stamping reports.

"If you work with the man, it's obvious that he must have had a primary research role, in all of the initial breakthroughs at least. Not every sample we grow comes out the way we want it to; in fact, most of them don't. Once you've been doing it for a while, you can just look at the wafer, and without doing any real sophisticated characterization, you can have this very gut feel of, This is probably what's wrong. In the early days, Shuji would do that: he would look at the sample under the microscope and say, Your nucleation layer is too thin. To kind of have that intuitive feel for it, you actually have to have done it. You can't have looked at somebody's plot of x versus y, and be able to walk into a research lab and do it.

"Or, when we would be testing our first LEDs, you'd always want to see them lit up. We'd joke that you need a calibrated eye: you know, we'd light it up first without having a spectrometer on it. Shuji would guess the wavelength—Oh, that's about 430! And he would usually be right, within plus or minus ten nanometers, whereas I would usually be completely off."

Shuji has also initiated his students into some of tricks of the device maker's trade. "For example, cleaving sapphire. We're always a little antsy about that," Kaeding said, "but he showed us how to do it. We typically grow on a two-inch wafer, and you need to break it down into pieces. And sapphire, it's basically a ceramic, I mean, it's a really tough material. If you've ever been around silicon wafers, if you drop them ever so lightly, they'll shatter. Gallium arsenide, indium phosphide are even worse: if you breathe on them incorrectly, they'll break into a million pieces. Sapphire, on the other hand, you can drop it from chest high onto the floor, and it'll bounce a couple of times. Because of that, sometimes when you're trying to break it into pieces, it can actually be hard to do. So when you break it, like the way you break glass, you run a scribe across it. And if you push too hard, or unevenly, you get these jagged lines, and you can sometimes get it to break into a bunch of pieces.

"Shuji was the one who's like, You just make a small mark, so that way you don't really damage it, pick it up with some wipes so you don't get it dirty, and just break it. I remember the first time, I was horrified. And he was like, Don't worry—once you've done a few thousand [times], it'll be very easy. After doing it ten times it *was* very easy, and it was a much more reproducible way of doing it than what I had been shown before. But that's another perfect example. It's something that it's obvious that he had physically done a bunch of times."

"Another thing he showed us," Sharma added, "was how to do quick tests on our devices. We assumed initially that you had to slap on your contacts before you could get any feedback, and that's usually at least half a day's work. He showed us a way to get feedback in essentially five minutes after the growth. Things like that, you know that the man's been in the lab himself and tried out these things."

Over the years, an excellent working relationship has developed between Shuji and his students. They respect him, but they are not intimidated by him. Seeing this easygoing relationship, visiting researchers from Japanese companies would comment that this was very different from the deference they were accustomed to seeing students according their professors back home. It seemed to them that Nakamura was not Japanese. Sharma knew what they meant, but he also felt that "there are aspects to Shuji that are very Japanese. Especially how hard he works. You can tell he's always thinking about work, because it might be 6 PM on a Sunday evening when you get a call from him, and he's got an idea for you to try."

"There really are no working hours and play hours with Shuji," Kaeding agreed. "I mean, we've had phone calls from him on a Saturday saying, Why aren't you in the lab? I'm in the lab, where are you? I want to talk to you. But at the same time, you get the impression that it's not so much like he feels required to work long hours, or every weekend, as much as, he doesn't really want to do anything else. I mean, we've asked him if he's got any hobbies—all Japanese guys play golf—and he says, No, he doesn't do any of that."

*   *   *

Soon after getting his new equipment up and running, Shuji decided to tackle a vexatious problem to which all nitride-based light emitting devices are subject. Normally, gallium nitride grows as a polar molecule, the atoms stacking so that there is a gallium face and a nitrogen face. Polar molecules carry opposite electrical charges at each end. Polarization is one of the big differences between the nitrides and other families of compound semiconductor materials. In an LED, polarization decreases the likelihood of electrons and holes combining to produce light. It is thought that the particles are pulled apart by a formidable-sounding phenomenon known as the "quantum Stark confinement effect." Nakamura realized that, by developing an improved method for producing nonpolar materials, he might be able to vanquish the Stark effect. Jim Speck and Steve DenBaars had already begun research on nonpolar materials at UCSB. Now Shuji brought his unparalleled expertise in MOCVD to the party.

Over the years many people, starting way back in the late 1960s with Herb Maruska at RCA, had tried to grow nonpolar material. Despite all their efforts, however, the quality of the material was always inferior to the polar equivalents. The issues were the lack of a suitable substrate, and an insufficient understanding of the growth chemistry. Now, by taking a radically different approach to the growth chemistry, Shuji and his students figured out how to grow nonpolar GaN films of unprecedented smoothness. Smooth surfaces being a prerequisite for growing devices, the inability to grow such films was the reason that no one had been able to produce nonpolar LEDs.

In 2002, not long after experiments at the lab got under way, Ben Haskell made a breakthrough. Under his professor's guidance and drawing on a colleague's MOCVD results, he employed Maruska's old method of hydride (aka halide) vapor phase epitaxy to grow smooth nonpolar GaN films. Following Maruska, he used a novel substrate material (lithium aluminate) to grow nonpolar material with areas large enough to fabricate working devices. By careful control of the growth parameters, Ben managed to overcome most of the problems that had plagued earlier efforts, consistently producing films of low-defect-density gallium nitride. In early 2005 Nakamura announced the first bright blue LED grown on these new nonpolar substrates. Though not as bright as polar devices, it was far brighter than any previous nonpolar LED.

Nonpolar material holds the potential to greatly enhance the efficiency of light output. Haskell anticipates that nonpolar material will reap the most significant benefits in the green and even the yellow parts of the spectrum, where polarization fields kill devices. So far, however, most of the effort has gone into blue. Nonpolar material (and a variant known as semipolar) quickly became one of the major research themes in Shuji's lab. The research has blossomed to become the kind of mutually supportive effort for which UCSB is noted. It spans several labs, with DenBaars, Speck, and their students also joining in.

"We're seeing this fantastic cross-collaboration," Haskell enthused. "The MBE guys will grow templates on one type of substrate, then they'll give it to me, and I'll use HVPE to grow a layer that uses lateral epitaxial overgrowth to improve the material, then I'll give that to the MOCVD guys, and they'll grow devices on it. I don't think you see that kind of cross-collaborative study where three different labs and growth techniques will be touching on the same sample during its life in too many places. Shuji's been extremely supportive of that kind of approach. And it's really paid great benefits for the quality of research that's going on."

"We're constantly improving the material quality," Paul Fini added. "Shuji likens it to the situation when he was at Nichia in 1992, right on the cusp of making his first bright blue LED. He had a billion defects per square centimeter, but the device worked, it just wasn't that great. But over the next four or five years, it improved incredibly. We're seeing the

same kind of progress in nonpolar. It's still in its infancy, but we antici-
pate significant increases in output power in the next couple of years."

The question of commercialization inevitably arises. From the time
he arrived at UCSB, it has been Nakamura's dream—his *American
dream*, he calls it—to have some of his students start up a venture business based on the results of research done in his lab. "We do think that
nonpolar has the potential to be truly competitive with what's already out
there," Fini said. Though their company is still at the conceptual stage,
Shuji's students have already come up with a name for it: Inlustra. In
2005 the business plan the pair of grad students drew up won first prize
in the annual competition run by the university's technology management
program.

*   *   *

Given his stellar track record, expectations have been high that Nakamura will go on making breakthroughs. Skeptics ask, What has come out
of UCSB since his arrival? Shuji's colleague, Jim Speck, has a good
answer for this question.

"Has Shuji done anything since coming to Santa Barbara that
matches the blue and white revolution he created? No. But has he done
things, which have created new knowledge, inspired students, and got us
all excited? Yes—this whole area of nonpolar LEDs is causing a huge
effect. If you look at the literature, you'll see tons of people doing work
in this area, because it has such great potential. For example, it makes a
huge difference for things like [LCD] backlights, because the backlights
used in laptops need polarized light, and this thing comes out polar, so
that's a huge advantage.

"So the man is not resting on his laurels, that's for sure. When is he
going to make his next breakthrough? God only knows.* But has he contributed mightily to us? Of course he has. I mean, look at the number of
students getting into the field because of him being here. These are met-
rics which are far more important than him doing anything else. I'm sure

---

*In late January 2007, UCSB announced "a major breakthrough in laser diode development" by a
team of researchers led by Shuji Nakamura—a nonpolar, low-power blue-violet laser.

he will, given time. Already he is making great strides. But he is inspiring a whole generation, and what more can you ask?"

In a sense, it would not matter if Shuji never made another breakthrough. He has already done enough. After all, as his old friend and sometime coauthor Gerhard Fasol pointed out, "Nakamura changed the world, no?"

Nakamura changed the world, yes—and for the better. Now, in the final chapters of this book, let us leave Shuji and move on to look at how the solid-state lighting revolution that he launched is getting under way. Not surprisingly, here too the leaders are not long-established lightbulb and lighting fixture manufacturers but aggressive, entrepreneur-driven start-ups.

Part Four

THE END OF EDISON

CHAPTER 14

In October 2005 I flew to Vancouver to see with my own eyes the ultimate lamp. I found it, in the conference room of a company called TIR Systems, made manifest in the form of a slim silvery lighting fixture suspended a few inches above a single red rose. The point of this elegant demonstration was simple: LEDs do not radiate enough heat to wilt the flower. Thus, restaurant owners would not need to replace the blooms on their tables twice a day (or, more likely, substitute plastic flowers for real ones).

But there is more than coolness to this light, dubbed the Lexel, for "light emitting pixel," and there are more applications for it than enhancing restaurant ambience. For the Lexel is arguably a perfect white light source: needing no warm-up, it comes on instantly, and stays on without flicker or other instabilities. Its beam can be changed from narrow to wide simply by inserting a low-cost plastic lens, unlike the expensive and crack-prone glass optics that conventional lamps need to deal with incandescent heat. The Lexel can be set to any color temperature, from warm (reddish) to cool (bluish). Remarkably—and this is something that no other light can do—it can hold exactly that color as you dim the fixture.

The secret of the Lexel is not the clusters of red-green-blue LED chips it contains, nor the feedback mechanism used to control them, fiendishly clever though this is, nor the 1,000 lumens of light that in its most powerful version the lamp produces. Rather, it is the fact that the design team at TIR Systems has essentially reinvented the lightbulb. In so doing, they have developed a universal lamp that suits the sophisticated needs of the twenty-first century. Will the Lexel replace Edison's lightbulb? It is still far too early to tell. What can be said is that the folks at TIR Systems have set the bar pretty high.

*   *   *

Unlike most of the other companies we have encountered thus far, TIR Systems was not specifically established to exploit the new technology of solid-state lighting. In fact, the company was founded in 1983, as a spin-off from the University of British Columbia, to commercialize "light pipes," transparent hollow tubes that can be lit up from within by a light source to provide a uniform light of any color. TIR Systems (the initials stand for "Total Internal Reflection") builds luminaires—lighting fixtures—for industrial and specialty markets. Light pipes are used in tough outdoor environments such as tunnels and for picking out the tops of skyscrapers.

At a lighting industry trade fair in 1997 Brent York, TIR Systems' chief technology officer, was approached by a friend from Hewlett-Packard. "He showed me one of their high-brightness red LEDs. This was after-hours, secretly. We took that LED and attached it to a four-inch-diameter, eight-foot-long light pipe at our booth. That one diode lit up the whole pipe, you could see it clearly across the floor." It was the pivotal moment in York's professional life. "My perspective from then on was that solid-state lighting was real; it was going to happen, there was absolutely zero doubt in my mind," York told me. "LEDs were going to change everything, and we as a company needed to be on the forefront of that."

One of the first things that LEDs changed for TIR Systems was the identity signage the company produced for corporations. In particular, the signature thin green strip, visible from miles away, that identifies BP gas stations. Replacing high-voltage neon with low-power LED-pumped light pipes at each of more than five thousand sites enabled TIR Systems to lop a total of 25 megawatts off BP's electricity requirements. "That's enough to power a small town," York said. At the same time, the experience of installing an average of two hundred feet of pipes per station, at twenty-four LEDs per foot, taught TIR Systems a lot about working with solid-state lighting. Notably, that when things go wrong, it is seldom the LEDs that are to blame. "In our experience, if you took all the LED chips that failed, they wouldn't even fill a thimble," York said. "But if you took all the power supplies that failed, they would fill two or three trucks. So the thing to understand about LED lighting is that, provided the LEDs are treated correctly, their reliability—measured by the number of dropouts and failures—is very, very high."

From individual colors, TIR Systems moved on to experiment with LEDs for color-shifting. The company became one of Color Kinetics' first customers, adapting the Boston firm's modules to work with its light pipes in large-scale applications. But illuminating the facades of buildings is just a niche. York was impatient to leverage the company's experience with LEDs to address much larger markets.

*   *   *

Brent York comes across as a thoughtful, deeply sincere person. He is passionate about lighting, a field in which he has worked for more than twenty years. York joined TIR Systems in 1985 with degrees in engineering physics from the University of British Columbia and Simon Fraser University. Turns out that engineering physics is a good qualification for solid-state lighting. "You have to understand something about multidisciplinary activity, about power, drive, analog, digital, optics, light, semiconductors, solid-state physics—you've got to have it all or you're not going to grasp this field."

Was it coincidence, I wondered, that Canadians seem to have embraced the technology of light emitting diodes with particular alacrity, what with Carmanah Technologies in Victoria, Light Up the World in

Calgary, and TIR Systems in Vancouver? "Not at all. I think it's completely by design," York replied. "You have to understand that Western Canada is environmentally conscious, globally aware, and a hotbed of technology. People want to live here—if you have to pick a place to live for technology and lifestyle, Western Canada, British Columbia in particular, is where it's at. It's probably one of the most beautiful places on the planet. We're just so close to nature here, you tend to be a bit more attuned to the environment. So technologies like solar power, fuel cells, alternative fuels, wireless, bio-resources—all this stuff is going on here." TIR Systems prides itself on being environment friendly. "We're one of the few companies in the world that is actually net-energy negative, that is, we save more energy than we consume. Carmanah can make the same claim. I think that's really important, it means a lot to us here."

Entrepreneurial culture seems to thrive in the clean air of BC. "I think entrepreneurship comes naturally to people who migrate to the West Coast. I have more in common with my Californian friends than I do with my Eastern Canadian friends. California, Oregon, Washington, British Columbia—we all share a common bond. We're called Cascadia for reason."* Pending the secession of the western states, however, York prefers to call TIR Systems a "North American" rather than a Canadian company.

*   *   *

Lighting has evolved over centuries. It is built on traditional forms. Walk into the lighting section of a Lowe's hardware store and you will see candelabras and carriage lamps, fixtures that harken back to an age when the sources of light were candles or gas flames. For as long as anyone in the lighting game could remember, lighting had been an extremely stable business, with little difference from one year to the next. Product life cycles tended to be measured in decades. Then along came light emitting diodes, and it was kiss good-bye to this cozy world. The ultra-staid lighting industry is struggling to adjust to the new reality. Things simply will not stay still.

---

* In fact, the proposed Republic of Cascadia does not include California.

"Solid-state lighting is literally flailing around, like a hose that you're gripping five feet from the end," Brent York said. "What you have is a light source that is changing from one month to the next. You've also got a packaging paradigm which is evolving all the time." Making life even more complicated for the already-befuddled luminaire makers is an almost total lack of standards. In order to specify requirements precisely, there need to be objective quantifiable measures for things like luminous output and lifetime. But LEDs are different from conventional light sources. The lack of standards also applies to components. "If you are a lighting fixture manufacturer and yet another announcement from Cree or LumiLEDs comes out, you shudder and go, You mean I have to change my driver? But I don't have the expertise to do that!"

"Before, all the lighting fixture guys had to do was bend some metal, plug it into the wall, and walk away, because the Edison bulb took care of the rest," commented Chris James, a vice president of marketing at Cree. "Now, all of a sudden, they have to be electrical engineers, mechanical engineers, thermodynamic engineers, optical engineers, and they don't even know what those words mean." The world's largest luminaire manufacturers are not accustomed to doing much research; they typically invest less than 1 percent of their revenues in R&D. Attempting to form an amalgam out of the technology-focused solid-state lighting business and the application-oriented lamp-making business is thus like trying to mix oil and water. Talk about worlds colliding.

York first noticed this clash of cultures at the same trade fair where he encountered high-brightness LEDs. "I remember seeing the LED guys there, the chip-heads as we call them. They had this tiny ten-by-ten booth with a saggy little sign that said 'Lighting from Hewlett-Packard.' They were standing around, scratching their heads, trying to understand this world of lighting. For their part, the lighting people were walking by, poking at these LED lamps and going, Gee, these are bright—where am I going to get them with an Edison socket? Both parties were like deer caught in the headlights."

In the interim, the mismatch had if anything gotten worse. Seeking to impose some order on the chaos, York came up with a clever idea. He would make it easy for luminaire manufacturers to adopt the new technology of solid-state lighting by solving all the technical problems for them. "It was brutally simple: the lighting industry needs to have a stable model, something that bridges all of this change, reduces it down to something that a manufacturer can take from a box, put in a luminaire, turn it on, and it works."

In 2002 York approached his board of directors with a radical proposal: to develop a universal LED-based light that could replace the incandescent bulb, the compact fluorescent tube, and the halogen lamp. "If we can develop a single universal platform that can emulate any of the world's most popular lamps, then we've got a pretty powerful business model." It would be one lamp to rule them all.

Such a light would be a godsend for luminaire manufacturers unsure of what to do about LEDs. "A lot of them have been standing in the wings, looking for something that would allow them to adopt the technology easily, that wouldn't require them to develop all the necessary skill sets." To carry out the research, York and his team created a separate unit within the organization, a small team modeled on the Lockheed Skunk Works. They called it "Area 51," after the top-secret air base in Nevada where the US military tests new aircraft.

"If you look at a lamp and boil it down to its essential elements, there are only three things that matter: How much light? What beam pattern? and, What color temperature? [Chromaticity, in industry jargon.] Our goal was to build a platform that could accommodate those three requirements, plus one other important consideration—volume," York explained. "You've got to be able to sell lots of something in order to create a low-cost model. So the Lexel was designed to take advantage of integration technology and to be manufactured on high-speed production lines."

York and his team began by going right back to basics, working on ways of dealing with the raw LED chips straight from the wafer breaker. As we have seen, there can be considerable variance in the performance of chips even from the same wafer. To compensate for this variation, chip makers sort their products into different bins. TIR Systems' engineers came up with a feedback loop that monitors the output of individual LEDs, altering their drive currents to iron out any differences in performance among them. That means the Lexel can accommodate not only chips with varying performance from the same maker, but also chips from different makers in the same fixture, for a significant saving in cost.

Thermal management—cooling—was another key issue that had to be dealt with. Though LEDs do not emit heat, they do get hot internally, especially when you push them hard by running large amounts of current through them. So the engineers had to come up with an efficient way of removing the heat from the system.

The most important consideration was the nature of the white light source itself. Ideally, this should hit a point on a curve derived from a theoretical construct called a "black body." Think of this as a piece of metal that, as it heats up, changes color from yellow to red to blue-white. That explains why light is specified in terms of "color temperature," measured in degrees on the Kelvin scale. The higher the number, the cooler—i.e., bluer—the shade. But lights other than incandescents, whose filaments are metal, typically do not output a continuous spectrum. Makers of compact fluorescent lamps go to great lengths to develop phosphors that approximate a point on the black-body curve. Most of the time they fail, landing above or below it, with the result that the light fluorescent lamps produce looks wrong, with a greenish or mauve tint.

TIR Systems' first attempts were the equivalent of a fluorescent lamp. "The light was pretty good," York said, "but I would walk into a room and though I couldn't put my finger on it, I knew that it wasn't exactly right." His team dug in, kept working away at the problem, tweaking the algorithm software, adjusting the feedback-loop hardware. Finally, after three years of experiments, they nailed it. One day a researcher burst into York's office. "He said, Brent, you've got to come and look at this! I walked into the room, which was solely lit with a Lexel, and I said, My God—it's perfect!" It was a compelling moment: The lamp had hit the black-body curve exactly on the mark. You could look at the light and not know what its source was. Now lighting designers who specified any given color temperature could be confident that, with the Lexel, they would get precisely what they ordered.

✳   ✳   ✳

There are two types of lighting fixture available. The more familiar type for most consumers are the track systems you can buy at Home Depot or some other big-box do-it-yourself store for sixty bucks. Then there is the type that commercial premises such as offices, stores, and restaurants use. These fixtures may cost ten times as much, but they are designed to last for up to fifteen years of continuous use. This second type is known as "specification-grade," or "spec-grade" for short. It was spec-grade lighting that TIR Systems chose to target initially. This made sense: the company had always been in the specialty lighting game; moreover, that is where most of the early adopters were. Commercial lighting was also where the opportunities to add the highest value lie. Initial prices in the United States for Lexel-enabled luminaires ranged from $750 to $1,100.

Historical parallels support this strategy. The halogen down lamps now commonly used as ceiling lights in residential housing began life in the 1950s, as wingtip lights in supersonic jet aircraft. From there, halogens migrated to the projector lamps in slide carousels. Then, in the early 1980s, halogens were packaged in reflective cups to adapt them for use in fixtures. Initially, these mirror-coated glass housings cost a lot to produce, making halogen lamps too expensive for most people to afford. Over twenty years, however, thanks to economies of scale, the cost of making halogen bulbs and reflectors has plummeted, with the result that they are now ubiquitous in homes. York believes that LED fixtures will follow a similar trajectory.

The Lexel made its debut in April 2005 in Las Vegas at LightFair, the US lighting industry's showcase. The pendant version of the lamp took pride of place at the TIR Systems booth. It shone with a cool light, the rose beneath it never wilting throughout the three days of the show. The lamp produced a high-contrast sparkle—what lighting people call "pop"—in the objects it illuminated. The Lexel attracted considerable interest from makers of spec-grade fixtures. TIR Systems subsequently signed agreements with four major lighting firms. They are Genlyte Thomas, which has a 14 percent share of the US lighting market, Lighting Services, a specialist in high-end lighting for the presentation cases used in museums and retail displays, and two large European firms, Spectral and Zumtobel.

Teaming up with big lighting companies was a smart move, thought industry observer Bob Steele of the market researcher Strategies Unlimited in Mountain View, California. "Concentrating on making the light engine, which can then be incorporated into these other companies' luminaires, will actually get them into the market. So for a small company with a great technology, I think that is the best path. TIR Systems did a beautiful job of engineering and integration, the Lexel is just a tour de force."

In February 2006, after just four months of joint development, Lighting Services introduced the first fixture based on the Lexel. Two months later, TIR Systems announced a major restructure to focus on its light engine. The company suspended development of its other lines, laying off many of its manufacturing workers and incurring big losses in the process. "Basically what they've done," said Steele, "is bet the company on the Lexel. If the Lexel succeeds, they'll do well, and if it doesn't, they'll go down in flames."

TIR Systems was also having to spend hundreds of thousands of dollars on patent-related legal battles with Color Kinetics. The distraction and waste of resources irritated York. "Look at Asian countries like Japan, Taiwan, and China, how quickly they are adopting LEDs, how much money they are committing to solid-state lighting. Meanwhile, we in North America are throwing stones at each other. We spend too much time squabbling over what is about to become the single biggest market opportunity in the history of lighting. If we don't watch out, our Asian friends are going to own this business. From my perspective, there's way too much opportunity, it's going to be shared by many of us around the world, so let's not be fighting over it at this point—we have a business to run!"

In May 2006 the Lexel featured prominently at Light + Building, an enormous trade show held biennially in Frankfurt that dwarfs LightFair, its American counterpart. Compared with US lighting companies, the Europeans appeared much more aggressive in their adoption of LEDs. TIR Systems' pride and joy was prominently displayed in the booths of Spectral and Zumtobel. Indeed, Zumtobel has gone so far as to create an entire subsidiary, Ledon Lighting, to develop Lexel-based products for the parent company and other customers. At Light + Building, the Austrian firm was showing a very cool-looking LED spotlight with the unusual name of Tempura.

Spectral exhibited Lexel-based luminaires that were intended for office environments. The emphasis was on adjusting the color temperature of the LED light according to the time of day or the season. The company reportedly believes that LED light can be used to affect people in a positive manner, promoting mental well-being and stimulating productivity. "The Lexel got a great reception in the lighting community," said Steele, who attended Light + Building. "In the Spectral booth, they didn't advertise that the luminaire was using LEDs. Lighting people were coming into the booth saying, What's in that? They were trying to understand how it worked. They were just amazed."

The Lexel applied to spec-grade lighting was just the beginning for LED-based fixtures, York thought. "Solid-state lighting is an industry in its infancy. The technologies still have a lot of evolution to go, there's more ahead of us than behind us, so the story is very far from over—in fact, we've barely got to our knees. I think we have to recognize how slowly the lighting industry changes; I'd be nuts if I said that I think it's going to change overnight. What I can say is, I think the wedge has been rammed into the door. Some of the stuff we're working on, oh man, it makes me sweat just thinking about it. I mean the things we have on the horizon, if you start looking at their true implications from a global perspective, you begin to realize that we really are right on that edge."

Consumers tend to follow an adoption curve where we see something in a commercial installation, then desire it for our homes. In North America, a recent example of this phenomenon is the vogue for home theater systems, which builders are increasingly offering as an optional package to add value to the houses they sell. Such systems may cost $10,000, but that is just a drop in the bucket for a $500,000-plus home.

York believes that lighting systems have the same potential to add value. "We're going to start seeing solid-state lighting in the home in certain key environments. A light source that is adaptable from a color temperature point of view would be enormously valuable in, say, the bathroom. I can see companies selling complete bathroom lighting systems to contractors. At first, they only go into the best homes but pretty soon the cost drops, the demand rises, and the average middle-class home starts to adopt this sort of bathroom system. If a female member of the family wants to take a nice bath, she can change the lighting to something intimate. If she wants to put her makeup on and have it look right for daytime or nighttime, it's automatic, it can be done, it just has a dial on the wall. So you buy the whole system, the contractor puts it in, and lots of people in the value chain make money. That's how it adopts."

Further down the track, York noted the growing trend toward localized power in the form of homes producing their own electricity. In a typical home, lighting consumes around 3,000 watts. Installing solid-state lighting would reduce power consumption by an order of magnitude. "Can I support 300 watts from a solar panel? You bet I can. Can I be autonomous even here in Vancouver? Absolutely—it doesn't take a lot to generate the kind of power required to light our homes. If I could do the whole thing in one, that is, operate my lighting system with a small solar panel and battery backup, that would be fabulous. I actually think Carmanah is on the right track here by offering systems that are solar powered. My personal dream is to get homes to the point where they only need 300 watts for all of their lighting, then move that off the grid, have that as autonomous, as part of the operating system of the house. To me, that would be a huge motivator. At the current cost of electricity, the payback on that could be extraordinarily short—less than three years."

Who will be first to implement LED lighting on a grand scale? Where is the bleeding edge of the solid-state lighting revolution? The answer is, places where energy efficiency is at a premium. And, as we shall see in the last two chapters of this book, few places are more conscious of the need for energy efficiency than the People's Republic of China and the State of California.

CHAPTER 15

When a revolutionary new technology emerges, the reasonable assumption is that it will be adopted en masse in the country of its origin or, at least, somewhere in the developed world. With solid-state lighting, however, the technology is being adopted on a grand scale first in a developing country, namely, China.

This unexpected development is the result of a convergence of circumstances. First, China's appetite for energy is growing too fast for its electric power generation capacity to accommodate. This is forcing the Chinese authorities to seek radical new ways to contain or reduce consumption. Second, China is in the midst of a building boom, with new construction going up across the country at an unprecedented rate. Such a fast-changing scenario makes it easier to adopt a new technology. Third, the Chinese are looking for ways to move from manufacturing simple to higher value-added, upscale products. In a new industry like solid-state lighting, the barriers to entry are lower. Fourth, to a far greater extent than would be possible in a noncommunist country, China can enforce the implementation of new technology by government fiat.

China is just one of several Asian countries that have organized large-scale national initiatives to promote the development of solid-state

271

lighting. Others include Japan, Korea, and Taiwan. The European Union also has various LED-related projects under way. The United States, by contrast, as we saw in chapter 2, has been slow off the mark. The federal-government-funded Next Generation Lighting Initiative was slated to commence in 2007. In addition to having a head start, unlike their American counterpart, the Asian initiatives are highly organized. For example, the Japanese LED Association, which has over seventy members, has produced detailed road maps. The association is promoting widespread public understanding of LED technology and its energy-saving potential.

China's Solid-State Lighting Program was founded in June 2003 by the Chinese Ministry of Science and Technology with support from six other ministries and eleven local governments. Involving fifteen national research institutes and more than fifty commercial firms, the R&D program had initial government funding of around $17 million. Its goal is to develop key technologies, applications, and markets for the nascent industry. Short term, its objective is colored specialty lighting; long term, white lighting for the general illumination market.

The Chinese LED initiative is directed by a dynamic woman named Wu Ling. A former doctor, Wu spent six years working in the field of public health as an assistant professor at a medical university. In 1989 some of her students got caught up in the Tiananmen Square protests, and she was implicated. A period of exile in the provinces followed. Then Wu went abroad, to Canada, where she got an MBA. She switched careers from medicine to consulting and venture capital, with a focus on new materials. "At the beginning of 2003, we decided that gallium nitride LEDs had a big future," Wu told me. "We did some study, then we made a proposal to the Ministry of Science and Technology. They offered me this job developing solid-state lighting strategy for the government."

Having a private consultancy lead the way on such an important project is an extraordinary model for a communist country to follow. But Wu Ling is not an ordinary person. "She's an incredible individual," said Robert Walker, CEO of Silicon Valley–based LED maker BridgeLux. Wu recruited Walker for her governmental advisory committee. "I've never met anybody with more energy and charisma and the ability to lead and mold people. Within six or twelve months, she became the face of solid-state lighting in China."

* * *

China's heavy reliance on coal to power its economy has already made it the world's second-largest producer, after the United States, of greenhouse gases. For the Chinese, the dismal air quality across much of their country is a constant reminder of this dependence on coal. In some places, you can almost taste the sulfur in the air. Inevitably, energy and its consequences for health and the environment are high on the Chinese political agenda. The government has set one of the world's most ambitious targets for energy conservation: to cut energy consumption by 20 percent over the next five years. At the same time, however, the country's leaders are determined that China's economy will quadruple in size by 2020. This will require at least a doubling of the energy supply.

Electricity generation is by far the biggest consumer of energy. According to Wu, 12 percent of electricity currently goes to lighting, about half as much as in the West. She estimates that if over the next ten years LEDs were to take 40 percent of China's lighting market, then the savings would be 100 billion kilowatt hours per year. That is more than the yearly output of the Three Gorges Dam, the world's largest power plant. Costing $24 billion, the giant project is expected to reach full operation in 2009.

"Faced with a great shortage of energy, the government will push solid-state lighting," Wu insisted. In China, when the mandarins want something to happen, they have all sorts of ways of making sure that it does. Top-down strategies include finance, both direct and indirect. Wu estimated that $725 million had thus far been pumped into the development of a domestic solid-state lighting industry. Some of this was private investment, but the bulk of the money had come, industry insiders believed, from government banks in the form of soft loans to LED-related start-ups.

Regulations—in both positive and negative forms—are another powerful lever. For example, officials can mandate that LEDs be used for cer-

tain applications, such as the illumination of well-known tourist land-
marks like Tiananmen Square. Or, authorities can turn a blind eye when
implementers cut corners. "China is different, in terms of regulations,"
said Bingwen Liang, CEO of Nanjing Handson Semiconductor Lighting,
a leading Chinese LED packager and fixture maker. "It's not so tight as
in Europe or the US. As long as you have good relationships with the
leaders, anything can happen. The leaders realize that energy saving is a
big deal in China, that's why LED lighting has found a great market
here." One leader who has got the message is Li Jian, the vice secretary-
general of the Chinese Ministry of Science and Technology. "Semiconductor lighting will inevitably replace traditional lighting all over the
world," Li said at a commissioning ceremony for an LED factory in
Dalian in March 2004, according to a report in the *China Daily*.

China is emerging as both a major market for, and a major supplier
of, solid-state lighting. Overall, the Chinese LED market is growing
annually at 40 percent in terms of units, at 23 percent in terms of sales.
It is serviced domestically by some six hundred firms, almost all of them
small outfits at the low end of the industry, which employ as many as
forty thousand workers to do labor-intensive jobs such as packaging
LED chips.

"The Chinese are going after solid-state lighting with a vengeance,"
said Bob Steele of Silicon Valley market researcher Strategies Unlimited.
"There's a huge, almost insatiable, market for LEDs. It's like a black hole
into which stuff just disappears." Nor is the market fussy about what sort
of LEDs it gets. "The off-spec, low-end stuff, gallium nitride LEDs
coming out of Taiwan that don't meet a certain voltage, or wavelength,
that stuff is sold by the kilo, literally—it's not priced by the chip, it's
priced by weight. It's just a way to get rid of scrap. The Chinese love colored lighting, they seem to have an infinite capacity to absorb whatever
is made. Some of it goes to make jewelry, those little blinking earrings
and necklaces. You see a lot of them. You walk into a bar in any Chinese
city, they've got the little blinking lights hung around the mirror, and they
are LEDs, just decorative lights."

Another unique factor that augurs well for the rapid take-up of solid-
state lighting technology is the fact that China's infrastructure is still a

work in progress. "The whole country is like a construction site," Wu said. In 2005 alone, Shanghai completed high-rise offices and apartments with more space than in all the office buildings in New York City put together. And that is just one city in a country that boasts more than 170 metropolises with more than a million inhabitants each. A regular visitor to Shanghai is Shuji Nakamura's colleague at UCSB, Steve DenBaars. "I'm very impressed with how fast they've built up the infrastructure there. The most impressive thing is how willing the Chinese are to try new technology. They may be the first ones to implement solid-state lighting on a broad scale, because they have government support."

In such a fast-changing environment, people tend to be more open to new approaches. "In the US, I'm not going to go to the Marriott Hotel and have them change their whole electrical system," Robert Walker explained. "In China, I get to go to a guy who's building a new hotel and say, Design your building around LEDs. There're a lot of greenfield sites in China, and that's a much easier sell. Because of all this new construction I can really take full advantage of the product. I think the adoption curve in the US is much slower."

In particular, the entire country is scrambling to prepare for the Beijing Olympics in 2008. These are billed as "the environment-friendly games" (of necessity: it would be embarrassing if the city's appalling air quality were to affect the athletes' health). LEDs are among the officially designated environmental technologies. The Olympics also represent a wonderful opportunity for China to show itself off to the world. One Taiwanese fixture manufacturer, NeoPac, has introduced a white LED lantern in the classic Chinese lotus flower design specifically for the games. But it is in colorful outdoor decorative lighting rather than general-purpose white illumination that the solid-state revolution is initially manifesting itself in the Middle Kingdom.

"The whole country's gearing up for the Olympics, not just Beijing. They expect a lot of foreign tourists all over China," said Steele. "They want to dress up the country, they see LED lighting as a key element in improving the appearance of buildings, bridges, fountains, and ancient sites." For example, Full Moon Tower, a 52-meter (170-foot) structure in Galaxy Park, Tianjin, is lit up at night by a dazzling computer-controlled colored light show. In the lead-up to the Olympics, city officials decided on an illuminated landmark to help raise the profile of Tianjin, a port city with ten million inhabitants located 125 miles southeast of Beijing. In urban centers across the country, it's the same story. "Every single local government is trying to do this kind of thing," said Xiao Guang He, executive vice president of Dalian Lumei Optoelectronics, China's second-largest LED maker. "If one lighting project turns out good, then people just keep on copying. The cost is now so low that people can afford it." Outdoor decorative lighting accounts for almost a quarter of the Chinese LED market, which was worth $1.4 billion in 2004.

There is a paradoxical consequence of this massive adoption of architectural-use LEDs. Namely that, so far from saving energy, the new applications are actually *increasing* the load on electric utilities. Fierce competition on price sometimes leads to corners being cut, with unfortunate results. "Decorative lighting is booming now, but if we don't take care of quality control," fretted He, "eventually everybody will just give up on LEDs. They'll think they are a bad thing."

Currently, the Chinese are relying on foreign suppliers to feed their massive appetite for solid-state lighting. Companies such as Cree and Nichia provide the chips, while color-shifting LED modules are supplied by firms like Color Kinetics and TIR Systems, which was responsible for lighting up Tianjin's Full Moon Tower. But the Chinese are determined to work their way up the value chain, to establish a strong industrial base for the manufacture of high-brightness gallium nitride LEDs. Already four local companies have begun to produce their own chips, with maybe twice that number gearing up to follow. Thus far, they have had little success, but they are not about to give up.

Xiao Guang He's company, Lumei, leads the field thanks largely to having acquired, for $9.6 million in 2003, the optoelectronic arm of AXT, a US firm based in Southern California. For the moment, Lumei continues to make its LEDs in California. But the technology is migrating inexorably eastward, carried to the mainland by returning Chinese-born, American-trained engineers. Like He himself, for example, who formerly worked for US laser maker Spectra-Physics, and Bingwen Liang, who spent seven years at HP-Agilent.

Though late starters, the Chinese draw strength from the fact that they are already the world's largest manufacturers of conventional lighting equipment. Also, from the success in solid-state lighting of their archrival, Taiwan. The island has rapidly made itself a force to be reckoned with in LEDs. "Five years ago, when I was telling people about Taiwan and LEDs, everyone pooh-poohed me," Robert Walker said. "Now nobody talks about this industry without mentioning Taiwan."

Taiwan already dominates global production of high-brightness red and amber LEDs, with a share of more than 80 percent. They now reportedly supply the majority of the world's blue LEDs (although, for the moment, they lag behind the Big Five in the key area of power chips used for illumination). Taiwan musters more MOCVD reactors than any other country. At one stage at least fifteen Taiwanese companies were manufacturing gallium nitride LEDs. Since then there has been some consolidation, through merger and acquisition. It seems likely that around three or four large Taiwanese companies will remain. Epistar, the island's leading producer of LEDs as the result of one such merger, is already bigger in revenue terms than Toyoda Gosei, the smallest of the Big Five. The main market for Taiwan's blue and white LEDs was formerly backlights for cell phone handsets. At the time of this writing, backlights for the LCD displays of personal computers and televisions are taking over as the market drivers.

For the moment, a wall of patents keeps Asian upstarts out of the lucrative Japanese-US-European market, forcing them to concentrate on their own home turf, that is, the Taiwanese-Korean-Chinese market. But as the Asians, especially the Taiwanese, develop their own intellectual property, and as the Big Five begin to license their patents, the wall is starting to crumble. Working on the principle of, if you can't beat them, use them, Nichia has taken a stake in Opto Tech, a leading Taiwanese LED maker. The Japanese firm also dropped its lawsuit against Epistar. Meanwhile, another of Taiwan's manufacturers, Edison Opto, has won government backing to develop high-power white LEDs.

In addition to generating IP, the Taiwanese also have a unique contribution of their own to make to the nascent solid-state lighting industry. Namely, their expertise in squeezing cost out of the manufacturing process. "That's not cheap labor," Walker explained. "It's a specific skill-set and a business-plan focus that they bring to the table. I think that's going to be a part of the industry, an important element in its success. You need to leverage the global talent pool. If you don't respect invention in the US and Japan, if you don't take advantage of the Taiwanese skills in terms of manufacturing and the European skills in taking the technology and designing beautiful new lamps and fixtures, if you don't bring all that stuff together, then the industry won't succeed as it should."

The Chinese are determined that anything the Taiwanese can do, they can do, too. Chinese makers are currently several years behind their Taiwanese counterparts, which in turn are still a couple of years behind the Big Five. Few analysts doubt that China, following in Taiwan's footsteps, will eventually become a significant player in the solid-state lighting industry.

But general illumination—in particular, residential lighting—will not happen first in China, because it is still far too expensive for most Chinese to afford. The solid-state lighting revolution will begin in California. Indeed, in the Eureka State, as we shall see in the final chapter, the substitution of LEDs for incandescent lightbulbs and compact fluorescent lamps in the home has already begun. And what happens in California today will likely happen across the rest of North America—and the world—tomorrow.

CHAPTER 16

How do you square the circle? You have a growing population, a slew of new electricity-guzzling multimedia gadgets like digital videos and plasma TVs, but for political, economic, and environmental reasons, you cannot build new power stations to service any significant increase in demand. So what do you do? If you are the state government of California, you pass laws. Via various carrots—such as rebates—and sticks—such as fines—they encourage citizens to do the right thing, energy-wise. The system works, wonderfully well: California's electricity usage is the lowest in the nation, by a considerable margin.* Amazingly, per-capita consumption in California is not growing; unlike in the rest of the United States, it is fairly flat. That is why other states look to California for their lead in setting energy efficiency standards.

But as California's population continues to inexorably increase, as people find ever-new ways to consume energy, and as generating capacity remains capped, the state's regulatory body—the California Energy

---

* 6,800 kilowatt-hours versus 12,800 kilowatt-hours for the US overall.

Commission—and its electric utilities must continually seek new ways to reduce consumption. The alternatives are unacceptable. You can spend a fortune to buy surplus hydropower at prime rates from the Pacific Northwestern states of Oregon and Washington. Or, worse, impose rolling blackouts because you cannot provide enough power, as happened during the California electricity crisis of 2000–2001, one of the most dramatic crises ever to afflict the state. "Blackout" means just what it says: the thing people hate most about power outages is that the lights go out.

Since lighting accounts for around a quarter of electricity usage, California has expended much effort attempting to get homeowners to switch from highly inefficient light sources like incandescents to more frugal forms of illumination. In October 2005 the state radically updated its code of regulations that mandate energy efficiency in residential buildings. Known as Title 24, the code already stipulated that "high efficacy" light sources must be used for, for example, 50 percent of lighting in kitchens and bathrooms. Now, the revision requires that the remaining parts of the house must also have high-efficacy forms of lighting. In many cases, that effectively means that the first light you turn on in your home must be a fluorescent. Or, more precisely, a compact fluorescent lamp.

CFLs are scaled-down U-shaped tubes grafted onto an Edison socket. Since their introduction in 1979, they have gradually gained acceptance. There is a lot to be said in their favor. Compact fluorescent lamps use up to 75 percent less energy than incandescent bulbs and last approximately ten times longer. Many of the problems traditionally associated with fluorescent lighting have been overcome. These included the characteristic delayed start and flickering as the lamp's ballast kicks in with the high voltage needed to initiate the gas discharge, the principle on which fluorescents work. Also, during operation, that all-too-familiar irritating hum or buzzing noise. Substituting electronic for magnetic ballasts took care of those.

But for all the lamp's advantages, most people have never learned to love compact fluorescents. For one thing, CFLs are expensive—about eight times the price of an incandescent bulb, three times as much as a halogen. Reducing the cost should be a simple matter of market forces and economies of scale. In the case of California, however, Title 24 has guaranteed the market for compact fluorescents. Manufacturers know this, and consequently are in no hurry to lower their prices.

For another, there is the ugly quality of the light that most CFLs produce, which many people find either too white or too blue. It is harsh—"eerie," some call it—unlike the warm yellow glow put out by an incandescent. You didn't get a true-color read from compact fluorescents, which meant that that nice granite countertop you splurged five thousand dollars on to upgrade your kitchen with looked a completely different color at night. Worse, they made people's skin look unhealthy. With their newer models lamp makers had fixed many of the color rendering issues, but negative perceptions persisted. As a result, only 6 percent of households use CFLs today.

A more intractable problem is that you can't dim compact fluorescent lamps. Turning down power to a CFL reduces the ability of its filament to emit electrons, with the result that it rapidly degrades. But not everybody was aware of this. What often happened was that buyers would move into a new home equipped with bright 26-watt fluorescents. After a week of living under harsh, stadium-grade lighting, they would dash off to Home Depot, buy a dimmer, hook it up, and blow the lights. Then they would call their builder and ask for replacements. This added time, labor, and cost. But for the builder there was also an additional issue. In getting rid of the unwanted fluorescents, builders now have to comply with another recent California law, Title 22, introduced in February 2006, which covers the disposal of hazardous waste.

Fluorescents contain mercury. Indeed, fluorescent tubes were originally known as "mercury vapor lamps." Mercury is an extremely dangerous substance, especially if it gets into the water supply or the food chain. Recent studies have linked mercury to autism in children and Alzheimer's in adults. There is controversy over this, but there can be no doubt that severe mercury poisoning does cause birth defects and paralysis. In Japan during the 1950s and 1960s, over nine hundred people died and almost three thousand were stricken as the result of eating fish containing harmful levels of mercury from Minamata Bay. Unfortunately for them, the bay happened to be where a local chemical company was dumping its contaminated wastes. Today, inevitably, many people fear exposure to mercury pollution.*

Home builders in California now need a permit to transport fluorescent lights. If the lamps break, they are required by law to dispose of them properly and risk a jail sentence if they fail to do so. For all of the above reasons, builders are not especially keen on fluorescent lighting. But obliged to comply with Title 24, what choice do they have? In fact, the code does not insist on CFLs. Any high-efficacy source that can meet the numbers it stipulates will do. And, as Dean Barnes, a builder of custom homes in Southern California discovered to his delight, an alternative light source had recently become available. Barnes would become one of the first builders in the world to enthusiastically embrace solid-state lighting.

*   *   *

Dean Barnes describes himself as a maverick. He has been on job sites more or less his whole life. He grew up in Seattle where his father was a general contractor, moving down to Southern California in 1970 when he was thirteen. Himself a licensed general contractor, Barnes has worked for the past nine years for Brookfield Southland. This is a semi-independent branch of Brookfield, one of the top twenty largest home builders in North America.†Barnes's outfit operates mostly in Orange County and Inland Empire, as the southeastern end of Greater Los Angeles is popularly known.

"We're a production builder in the semi-custom market, we run around 450 units a year. Our average sales price is probably in the low eights right now. We start in the low fives then go up to two million dollars. We don't get a lot of first-time buyers; our typical buyers are move-ups. They know what it's like to own a house, they know all the changes

---

* Compact fluorescent lamps contain only trace amounts of mercury, not large enough to pose a threat to householders, but sufficient to make their disposal at landfills and incinerators a concern. Ironically, fluorescent lamps can actually be seen as one of the best mechanisms for *reducing* the amount of mercury that enters the environment. The largest source of mercury pollution is coal, which naturally contains mercury. When coal is burned, as at coal-fired power plants, mercury is released into the atmosphere. Thus, switching from incandescents to an energy-efficient technology like compact fluorescent lamps reduces the amount of electricity required, hence coal burned and mercury released.

† The company is headquartered in Toronto.

they'd like to make from what they're living in now. We tend to allow our buyers to make minor changes if it doesn't mess up the schedule too bad. We try to accommodate them, keep everyone happy.

"My role is to specify material for option products. I investigate new technology home-building items, then massage them into packages that we can exhibit to buyers. I push a lot of new product into our option line, to make things available for homeowners. If it's kind of cutting-edge, Brookfield likes to be able to say that we offered it to the public first." One example of an option that Barnes had pioneered at Brookfield was "structured wiring," bundling all the TV and Internet cabling together in preparation for Internet usage throughout the home.

"Because of my custom side, I'm used to having people ask me, Why can't we do something cool? And that's kind of where this LED picture came to life." For homeowners, lighting is not a particularly big deal; it's something most of them take for granted. From a builder's point of view, the goal is to make home lighting feel natural, not draw attention to itself by not fitting in architecturally. At the builder expositions he attended, Barnes had seen numerous LED offerings, under-cabinet lighting, step lighting, accessory sort of things, but as yet none of the makers had attempted to accommodate fixture design into their products. "At the time I felt kind of locked in to fluorescent lighting, but it was something that everyone else was doing, and I didn't think that it was the most attractive form of light. I wanted to get into LEDs, which are smaller, they can be molded into something more pleasing to the eye, so we're using a smaller source but getting better light from it."

His chance came when, in mid-2005, Barnes was invited to a lunch-and-learn session at the offices of a small maker of LED signs and step lights called Permlight, located in nearby Tustin. There he met the company's CEO, Manuel Lynch. "I got together with him and I said, OK, you've got all these little LED modules that you use on your commercial signs—What if you redistributed them within a different case and offered them from a home buyer's perspective, would that be feasible? Then Manuel said those magic words that builders want to hear: We can do that for you."

*   *   *

He made a lasting impression on people, this Manuel Lynch. He was in his late thirties, over six feet tall, solidly built, with cropped dark-brown gray-flecked hair and a neatly trimmed mustache and chin beard. He evinced a determined, self-confident air. "Manuel's always got a smile on his face, and when you talk to him, he's got that excitement in his voice," Dean Barnes said. "He knows that he's on the cutting edge and he loves being there."

Deducing the origins of someone with a name like Manuel Lynch is not that difficult. His mother is from Spain, taught Spanish when she first arrived in the United States, later becoming a federal court interpreter. Manuel grew up in Southern California but spent his summer vacations in Spain and, like his siblings, is a fluent speaker of Spanish. His father is New York Irish, an engineer who spent his entire career at McDonnell Douglas, retiring as the company's head of aerodynamics. The motorcycles Manuel rides reflect his bi-continental background, an American V-Rod Harley and a European Vespa scooter.

Lynch entered the high-tech arena from an unlikely direction. At the University of Dallas, he majored in theater lighting. On graduating, he realized belatedly that, at least early in their careers, theater people tend to spend most of their time doing pick-up jobs like waiting tables. "I realized that I wasn't a really good waiter, so I decided to go into the semiconductor industry, because my brother was working there." It was while putting together a multimedia promotional video for his brother's company, Solid State Devices, that Manuel discovered his true vocation—marketing. But his theater degree courses had taught him one important lesson: "When you do a Shakespeare play that's based back in the 1500s, and you have to modernize it for the present day, you have to understand the history: What happened in the politics of the time, the economics of the time, what was the popular art like, what was the dress like? So any time I get involved in anything, I study it down to the tiniest detail. I find all the right people to help me understand it the best I can. I'm a quick study—I can study any new technology and get involved in it overnight. That's just my mindset, that's what I do."

From Solid State Devices, Manuel moved on to Microsemi, a Santa Ana-based chipmaker where he headed up business development. There, he was faced with the challenge of finding a common goal that would unite twenty disparate business units. The market he chose to address was electronic components for implantable medical devices like pacemakers. To learn more about this business, he identified every pacemaker company in the industry and asked them what they needed. The strategy paid off: the company increased the number of implantable components it made from two to seventeen. "My claim to fame is that I have seventeen products in Dick Cheney's heart," he joked. During his time at Microsemi, the company's stock soared from eight dollars to a high of eighty-five dollars a share.

Lynch first became interested in LEDs at Microsemi, identifying potential markets that could be pursued. The company was strong in thermal management technologies. Applied to light emitting diodes, this ability to prevent chips overheating internally meant that the devices could be driven harder. The initial target was brighter blue and white backlights for mobile applications such as cell phones and personal digital assistants. In early 2002, after a management shake-up, Lynch left Microsemi. Still aggressively pursuing LED applications, he began consulting for Permlight, which he joined as CEO in late 2002 when the company's founder retired.

Permlight was founded in 1995 by an engineer named Ben George. His goal was to replace incandescents with LEDs in the lights that are embedded in the floors of cinemas to prevent patrons tripping and falling. The company was successful, but the market was tiny. Permlight's engineers subsequently came up with the idea of adapting an LED product they had developed for theater seat lighting to replace neon in "channel letters." These make up the large illuminated signs you see on the outsides of commercial premises. They typically depict the corporate logo.

Permlight filed a unique patent that was based on the notion that mounting LEDs directly on the metal of the signs would be a wonderful way of dissipating heat. The big letters would function as massive heat sinks for the tiny lights. As with Microsemi's thermal management technologies, the advantage here was that you could run the LEDs using more than their rated current, making them shine brighter without hurting the chips. This ability is crucial because it allows Permlight to offer the most cost-effective products in the marketplace, since they use fewer LEDs than their competitors to achieve the same light output.

When Manuel Lynch arrived at Permlight, it was a stodgy old-style family business whose principals seemed more interested in playing golf than in making sales. There was talk of selling their technology to a Chinese company. The arrival of the dynamic young CEO—Lynch was then just thirty-five—soon put a stop to that. Within a couple of years he had completely restructured the company, reducing sales as a percentage of goods sold and doubling gross margins. To fund development, he tapped wealthy individual angel investors. Based on his previous track record, they were prepared to bet on the small, privately held company.

Lynch hired a team of like-minded individuals to pursue markets aggressively. The first was channel letters. By 2005 LEDs had replaced neon in almost two-thirds of applications. Permlight is the world leader in white lighting for solid-state signage, shipping miles of product every month. Customers include Ford, Sprint, and Chase Bank. Next, following the introduction by Nichia of powerful white LEDs that shone with a warm, incandescent-like light that worked well for natural skin tones, Permlight introduced a series of new products targeting the new-home market.

Lynch approached this market with typical thoroughness. He was determined to learn the builders' language, to crack their code of secret handshakes, to find out which golf tournaments to line up for, what the rebate programs for electricians were—whatever it took to get into residential applications. Accordingly, he had a list made of the 184 home building offices in California that purchased lighting products. He invited their buyers to a lunch-and-learn at Permlight's offices. "Dean Barnes was one of the people that came by, he asked tons of questions, and eventually he became a rock-solid customer. He's not by any stretch of the imagination a large customer, but he was one of those invaluable customers we identify who understands the trade, and the applications. He taught us about the dynamics of the way the business works, about who the distributors in the marketplace are, about the different trim sizes."*

---

* Though technically "trim" is just the visible part of a fixture, in builder's jargon, the terms "trim," "can," "can trim," or "recessed can" are used interchangeably to mean a ceiling-embedded lighting fixture.

Following that initial meeting, the two men spent so much time together that it seemed to Dean that he and Manuel were like old school buddies. "Manuel didn't know from a builder's point of view what he was up against," Barnes said. "I thought LED technology was a cool thing, a neat direction to go in, something that could be energy-efficient. If we got it designed in the right way from the start, redesigned the structure so that it was strictly a residential product, we could make it cost-effective as well."

Lynch took Barnes's advice. "What we've done is modularized our technology such that we have three basic systems that can fit pretty much any type of light source that's out there. They can be incorporated into every single fixture that goes into a house, whether it's a recessed can, an under-cabinet light, a pendant, a step light, an exterior lantern, or a closet light." Crucially, Permlight developed fixtures that were easy to install, with no messing around converting AC to DC. "A huge deal that nobody ever figured out is, Can electricians install these lights and not charge a premium?" Lynch said. "We developed systems that install just like fluorescents, and run on line voltage."

One good thing about LEDs was that, unlike fluorescents, you could dim them. Amazingly enough, as you dimmed them, their energy efficiency actually increased, so you got more lumens per watt. From a home builder's point of view, there was also another, less obvious attraction in switching from incandescents to LEDs. That is, the 80-percent reduction in electric power requirements for lighting means that builders can reduce the amount of copper cable they have to run. This is particularly beneficial in high-rise complexes that contain dozens of apartments, where the amount of copper cable is accordingly long.

For Barnes, however, getting approval from his management at Brookfield Southland to install Permlight fixtures was an uphill struggle. LEDs were new and unfamiliar and, in terms of initial cost, they were very expensive. "A builder's purchasing department is pretty much stuck on the bottom line," Barnes explained. "Their objective is to get something as cheaply as you can. When you tell them, This is a better product even though it's going to cost us more money, they don't necessarily welcome you with open arms. Because they can do the same for less, even though it may not look as pretty. Management doesn't consider LED lighting cost effective for the business plan, so I have to force it along.

"In terms of cost, you're looking at about fifteen dollars for a standard incandescent can trim. The going rate for Manuel's LED trim is about forty-five or fifty dollars. From a home builder's point of view, a 300 percent markup does not work on a one-to-one basis. So now you've got to start looking at it from an energy-efficient point of view, and playing on that marketing aspect of it. You go to a homeowner and say, You know what? If you buy this upgrade and put LEDs into your standard cans, which we have the ability to do, now you don't have to change a lightbulb for ten or twelve years. When you start figuring that you're going to pay a dollar for a lightbulb, and you're going to change it every six or eight months, you play that out over ten or twelve years, and the LED will pretty much pay for itself. In addition to which, when you get into the nuts and bolts of your utility bill, there are some savings there, it just takes a while to gain them." Predictably, Lynch was more upbeat. He estimated that the cost of upgrading to LEDs for an average US home, which uses fifteen to twenty recessed can lights, would be about $750. He claimed that the lights would pay for themselves in terms of savings on utility bills in eight to nine months.

The first homes Brookfield Southland equipped with significant numbers of LED lights were in Irvine, at a housing project named Treo. The challenge for Brookfield had been how to make its homes stand out from those of other builders. "Everything had to be different from the competition," a corporate marketing person said, "but it's differences by degree, not paradigm shifts. It's really about identifying a smaller niche of the market that demands more and is willing to pay for it."

At the time of this writing, Brookfield Southland had equipped around thirty homes in five different projects with Permlight fixtures. "We use them for under-counter lighting, for recessed can trims, a lot of artwork and bookcase illumination, hallway lighting, and lighting that is going to be on most of the time," Barnes said. Initial customer response was favorable. "Buyers like the look of it, it's got a nice clean look. From a decorator standpoint, going to LEDs for under-counter lights and being able to dim those lights in a kitchen or bar area means that you can create moods you can't do with fluorescents, and that's a big draw for those houses." Following this successful proof-of-concept stage, the company planned to incorporate LEDs into its standard specifications for all new homes.

Dean Barnes enjoyed being in the vanguard of the lighting revolution. For him, it was a new adventure: "Usually you go to the local building materials supply house, you buy what they offer, and that's what you have to use. To be able to have an outlet where you can go to somebody and say, Hey, what if we did this? Y'know, it's been an exciting time for me, working with Manuel, trying to be in the forefront of things, to be a leader not a follower."

Before going the whole hog and equipping entire homes with solid-state lighting, however, Barnes was looking forward to LEDs receiving an explicit endorsement from the California Energy Commission. This would take the form of a further update to Title 24, due in 2008, that would specifically list solid-state lighting as an acceptable high-efficacy source. "That's going to make a wide avenue for us to go down and say, OK, we're going to get rid of fluorescents, which are ugly and don't produce a good light, and change to LED trims throughout the house. Once the California Energy Commission officially announces that LEDs are as good as fluorescent lighting, it's going to open up a marketing opportunity for LED manufacturers and anybody that has the ability to design fixtures. I think that in 2008, this is going to explode on the market."

Already ten or fifteen high-end home builders in California were following Brookfield's lead. Manuel Lynch was confident that a precedent had been set on which other builders throughout the United States would quickly pick up. In May 2006 Permlight began offering recessed LED cans that complied with the high-efficacy requirements of Title 24. "We can confidently state that at least 50 percent of the top home builders in the USA will be using these products in new homes as the primary light source by the end of 2006," the irrepressible Lynch predicted. That year Permlight had thirty-two employees and sales of over $15 million. The company is growing at a rate of between 200 and 300 percent a year.

Most of the buyers would not come knocking on Permlight's door. Builders would buy direct from lighting fixture makers or from home building supply stores. Like Brent York of TIR Systems, Lynch had quickly recognized that "the general lighting market was very archaic, very old school. We thought if these companies were going to try and take LEDs and merge them into their world, they were going be really confused. They needed a partner that knew how to merge those LEDs into conventional lighting systems. That's what Permlight has become, an enabling engine for those companies."

In March 2006 Permlight announced the signing of an agreement with Progress Lighting. This Spartanburg, South Carolina–based firm is the leading supplier of residential lighting in North America, with almost a third of the market. Two months later, the partners announced the launch of more than fifty solid-state lighting fixtures. To augment the direct channel, Progress planned to sell a range of Permlight-based LED lighting products via one of their largest customers, Home Depot.

When would the residential market for LED lighting take off? Lynch was bullish in his predictions. "I look at the market, there's 1.2 million homes built in the US per year, and the average home uses 15 to 20 cans per house. Let's say we've got a potential of 22 million cans. Progress Lighting has a 30 percent market share, so let's take that down to 6.5 million. Let's assume that a good year would be half of 1 percent for them, switching over to LEDs. That's going to be 5,000 cans in the first year. That's a lot of dollars, and that's just for one part type, it doesn't include the step lights, or the under-counter lights. We looked at that, and asked, Is half of 1 percent realistic? And the head guy at Progress said, If we don't do 10 percent conversion in the first year, we're doing a bad job. So now we're going to be doing closer to 600,000 cans at 9 LEDs per, we'll need 5.4 million high-output white LEDs—that's a lot of sales. At $50 for an average per can—that's a lot of business.

"That's why we're starting to see it happen. If we reach 10 percent by year two—2007—we think that will be reasonable. We'll probably do around 5,000 to 10,000 with Progress Lighting of 10 different can types, for a total of 50,000 cans. In year two, we would expect those sales to go up pretty dramatically, as a result of Home Depot, which is our largest customer, to a market share of 20 to 25 percent. By 2008, when the rewritten Title 24 lighting standards come out, we expect that the states of Texas, Florida, Wisconsin, and New York will have adopted similar energy codes. We think the market share for LEDs in new construction could potentially grow to 50 to 60 percent by the year 2008."

Overoptimistic? Probably, but even if only a fraction of Lynch's predictions come true, they would still represent a giant leap for solid-state lighting. At the 2006 LightFair trade show held in Las Vegas in June 2006, Lynch said he was "overwhelmed" by the level of interest in Permlight's products. "Every major lighting company in the world dropped by to take a look at what our systems were and to discuss how we could integrate our lighting into theirs. They're actively pursuing us to do the same thing that we're doing with Progress." Many of these potential customers had been directed to Permlight from the Nichia booth. The relationship between the two companies was, Lynch liked to think, synergistic. "Nichia says that when they come to California, they visit Apple and Permlight, because Apple is the business for today [Nichia makes the LEDs for iPod backlights] and we are the business for tomorrow."

* * *

I was curious to know what motivates an entrepreneur like Manuel Lynch. "It's not about money," he insisted, "that's such a wrong focus. My whole life, anytime I decided to get involved in anything, it was like I grabbed a really big sword, I was going to fight any beast that was in the way, and I was going to win. It's about winning: when I chose a battle that I want to win, I'm not going to lose. That's what it comes down to, winning and winning and winning."

Winning is not merely a matter of personal satisfaction, though. "This is about winning major energy savings in the United States. I often get into trouble because in the presentations I give, I point out that it takes forty-two barrels of oil a year to light someone's home today with incandescent lighting. If you do it with LED lighting, it would take eight barrels. Nobody really understands that we can solve our national security issues and dependency on oil by saving energy. Gary Flamm, the head of the California Energy Commission, told me that if all homes in California were built today with incandescent lights, the lights wouldn't turn on. And that's been my big push—let's go conserve energy, we don't have to wait, let's do it today."

Lynch positively relishes the challenge. "I was at a lighting conference where a representative from one of the chip makers got up and said, LEDs are not going to be in people's homes until 2015. And I thought to myself, That's crazy. I'm the type of guy, if somebody tells me it can't be done, that's the first thing I'm going to do."

"Manuel's like a freight train," laughed Michael Siminovitch, "but it's zealots like him that make stuff happen. We desperately need to have these kinds of guys around."

Siminovitch heads the California Lighting Technology Center, a unique organization whose charter is to promote the application of energy-efficient lighting and to help innovative energy-saving technologies make the transition from laboratory to marketplace. CLTC is based at the University of California at Davis. This is a strategic location that is far enough away from the Bay Area and Los Angeles to be seen as independent, but close enough to the seat of government in Sacramento to influence the formulation of state energy policies. Established in 2004, the center works with its sponsors, the California Energy Commission, the state's utilities, and the electrical fixtures and components manufacturing industry.

CLTC musters a staff that includes specialists in electronics, optics, thermal management, solid-state lighting, and design. Its activities include doing research and demonstrating the technologies they develop for state buildings. The center also has teaching courses, for example, for builders who need guidance on how to meet the state's new energy codes and how to implement solid-state lighting. It also tests new products like Permlight's recessed cans, to see whether they meet Title 24 conditions for efficacy. In January 2006 Permlight joined the CLTC as an affiliate member. The company sponsors a class for students in the design of next-generation home lighting systems based on LEDs.

As an undergraduate, Siminovitch studied engineering in his native Canada (he hails from Ottawa). From there, he did a master's degree in industrial design at the University of Illinois. This was in the late 1970s, during the Carter administration, when the oil crises had made energy efficiency a hot topic. He met an engineering professor who was working on some early Department of Energy lighting standards. The idea that lighting could be a combination of science and art fascinated Siminovitch. It was a rich area, one where there was a lot to be done and where it was possible for an individual to make a difference.

From Illinois, he moved down to the Lawrence Berkeley National Laboratory in California. There, in 1997, Siminovitch made a splash by designing an energy-efficient, compact fluorescent torchiere fixture to replace 300-watt halogen torchieres. These had not only been responsible for a huge increase in residential electricity use, they had also caused numerous fires, notably in college dorms (where the lights were especially popular, students not having to pay the energy bills). Though the torchiere was a success with fixture makers and utilities, it was not the sort of thing that a national laboratory, whose focus has traditionally been on basic science and academic output, was comfortable with. For his part, Siminovitch was much happier filing patents than writing papers.

"What makes Michael different than any other scientist is that he figured out that, in order to be successful, he needed to get products out into the marketplace quickly," Manuel Lynch noted approvingly. "He's more of a real-world guy." Siminovitch was also, Lynch thought, the number-one authority on solid-state lighting in California. "He's got a really good sense of what's going to happen with LEDs, and how they're going to take off over the next couple of years."

I asked Siminovitch to give me his vision of the future of solid-state lighting, and its significance in the energy-efficient scheme of things. He began his answer with a caveat. "LEDs are important, but they're only going to be one of the arrows in the quiver." Improving the efficacy of lighting sources was only part of the solution. The other parts relied not so much on technology but on smarts, of which remarkably few have thus far been applied to lighting. For most people, lighting is an afterthought, not a principle concern. As a result, few homes and offices are lit as well as they might be.

Another arrow is control systems. "We've shown numerous times that simple lighting control systems are probably the single biggest opportunity for energy savings in this country." For example, a simple bathroom lighting system that Siminovitch and his team developed at UC Davis in conjunction with a local utility and two manufacturers. The system consists of a compact fluorescent light, which is triggered by a motion sensor, and an LED night-light. When the bathroom light is accidentally left on, and there has been no motion in the room for a given time, the main light turns off, leaving the backup LED light on. The night-light glows sufficiently brightly for residents to avoid tripping or walking into doors. The system is estimated to be 50 percent more efficient than standard bathroom lighting. It is expected to provide substantial energy savings, especially in hotels.

The LED revolution would happen, Siminovitch thought, in two ways. "One is an evolutionary process, through cannibalistic opportunities. That is, we take something out and put something else back in, an LED fixture that does a better job than what it's replacing. A good example of that is Manuel's business—he's doing pendants and downlights, consumer-grade products where the conventional lights are all 17 lumens per watt or less, and he's replacing those with 40-lumens-per-watt LEDs that last a lot longer. So it's a cannibalistic opportunity where you get some incremental benefit—higher efficiency, longer life, no mercury—but the fundamental job has not changed, you're essentially providing it as a downlight, something you recognize as a downlight.

"The cannibalistic opportunities are going to be great—anyplace where we've got an incandescent light source today, we're going to see an LED tomorrow. Incandescent sources are used predominantly in residential applications, so homes are a huge opportunity. Incandescents are also used in major amounts in merchandising, where people want tight beam control or spotlighting. So in terms of displacing load, eliminating mercury, reducing maintenance, all those nice things that really make a difference, the societal benefit is going to be tremendous.

"In addition, I think there's going to be a whole other area that's coming up, of revolutionary change, where we actually start doing new things in lighting because we've got this new technology. Things like spectral change, in which the colors change as a function of the time of day, or in terms of mood, or application. We know that very high color temperatures create certain levels of vision, and low color temperatures create certain levels of comfort. We have the ability now to do that dynamically, with control systems that were previously very expensive or very impractical to do with traditional light sources.

"The manner in which we light is very much driven by our technology. If you look at light sources, they have not changed that much. We've got an incandescent source which is relatively old and a fluorescent source which is relatively new, but the way we apply them is very primitive, we haven't even touched on it. So with the availability of electronic light sources, the capability of making both spatial and spectral chances relatively easily is going to open up new horizons. Coming down the road, it's going to be all the things we're going to see from LEDs that offer new ways of doing things, not just displacing what we've done before."

Somewhat to his own surprise, Siminovitch has himself recently become a user of solid-state lighting. "I'm putting LED downlights in my kitchen, because in my opinion, the technology has arrived and now it's only going to get better. I like the performance, I mean, at 40 lumens per watt, it's cost-comparable to traditional downlights, it doesn't cost appreciably more than the energy-efficient alternatives, I like the color, it's better looking than fluorescents, it looks like incandescents, and it has a lot of interesting control features. So I think the revolution is happening now."

Siminovitch is what diffusion-of-innovation theorists call an early adopter. If the number-one lighting authority in California, the number-one state in energy efficiency, is implementing LED fixtures in his own home, it will surely not be long before the rest of us—pragmatists, conservatives, even skeptics—follow suit. The solid-state revolution in lighting has begun. The circle has been squared.

CONCLUSION

T en! Nine! Eight! . . ." Historians chronicling the solid-state revolution in lighting who seek to pinpoint a watershed—one epochal moment when the era of the incandescent Edison lightbulb ended and the age of the light emitting diode began—could do worse than choose the stroke of midnight on New Year's Eve 2006. On that night, as usual, millions of revelers chanted the countdown as they watched the brightly lit crystal ball slide down the flagpole atop the One Times Square building in New York City.

Exactly one hundred years earlier, in 1906, when the ritual of dropping the ball to mark the arrival of the new year was enacted for the first time, the orb was lit with the then still-new technology of incandescent lighting. The most recent version of the six-foot-diameter sphere, introduced to mark the millennium, sparkled with 696 halogen lamps. Now, for its centennial, the symbolic ball was refitted with yet another new technology—light emitting diodes. The revolution in lighting that Shuji Nakamura sparked thirteen years earlier had reached critical mass.

There were plenty of other portents to which one could point. For example, 2007 would see the introduction of the first mass-produced cars to sport LED headlights. The Lexus LS 600 was equipped with white

LEDs, manufactured—appropriately enough—by Nichia. Nakamura's old firm was still leading the way in high-brightness LEDs, with a 22 percent share of the market. The company's new white light power chips were knocking the socks off everything else in the market. Customers rated these LEDs very highly, especially in the area of color consistency, a key consideration for lighting manufacturers.

The world's largest manufacturer of LEDs was reportedly targeting sales of $2 billion by 2008, of which automotive applications would account for 15 percent. Nichia's business plan also called for aggressive hiring, expanding its work force from 3,700 to 5,000. Old Nobuo Ogawa would have been delighted. Obviously, the company was getting along just fine without its former research ace. "Shuji trained some very capable people," was how one industry insider accounted for Nichia's continued prowess. At the same time, the company could now afford to hire the very best, including some top-notch American researchers.

Developments were happening fast, and the tempo was accelerating. The 100-lumen-per-watt barrier had recently been broken. What began as a trickle of products had turned into an unstoppable torrent. LEDs are turning up in almost every conceivable kind of lighting application. Another milestone was the introduction of a potentially iconic product by the furniture maker Herman Miller, of Aeron chair fame. Designed by Yves Béhar, the sleek Leaf lamp was a conscious attempt to take advantage of LED technology to produce a flexible, self-contained lighting system as a hip desk accessory.

Before it can fulfill its destiny of becoming the dominant technology, however, solid-state lighting still has a ways to go. Veteran industry observer Bob Steele of Strategies Unlimited estimated that in 2006, sales of high-brightness LEDs used for illumination were around $250 million, accounting for less than 2 percent of the world's $15 billion lighting market. Though backlights for mobile applications still dominated the market, profit margins for chip makers were much higher in illumination. Steele reckoned that the value of the illumination market would quadruple by 2010. "It's growing very fast," he said.

At a lighting industry trade fair held in mid-2006, Cree conducted a survey of more than a hundred exhibitors. Half reckoned that, by 2009, LED fixtures would account for more than 50 percent of their sales. Over 60 percent thought that solid-state lighting would begin replacing fluorescent lamps in offices and commercial premises within five years. That is, by 2011.

Of course, conventional light sources like incandescents and fluorescents will continue to hang around for many years to come. After all, we still use candles. But even in this oldest of lighting technologies, LEDs were making their presence felt. LED candles offered the same flickering aura of warm light, but without the messy wax, fumes, and danger of fire. Nonetheless, for Edison's lightbulb, the writing was on the wall. Incandescents are doomed to go the way of vinyl records, audio tapes, floppy disks, cathode ray tubes, and dial-up modems, all of which have long since been replaced in new applications by superior technology.

As with transistors, some advantages of the new technology are going to be related, not to how much energy LEDs save, but to the value that they bring: What you can do to your space, how unique you can make it look, and how much control you have over your lights. Would consumers go for a system that allowed them to control the color temperature of their room lighting to mimic some kind of circadian pattern? Or, to generate clouds of light floating across their ceilings? And if so, how much would they be prepared to pay for that?

The likelihood is that the paradigm shift will be quicker than most pundits have predicted. Certainly, the relentless focus by chip makers on improving the brightness of their LEDs is driving up the number of lumens output per watt much faster than anyone imagined. Now, the industry must turn its attention to the other curve in Haitz's law, the number of dollars per lumen. This is not dropping as rapidly as was hoped. In any consumer market, price is usually the greatest obstacle to widespread adoption. Not to worry, though: the semiconductor industry has heaps of experience in increasing yield and decreasing cost. Already some companies, like Robert Walker's BridgeLux, are focusing their efforts on this side of the equation. The lesson of history is, be optimistic.

One thing is certain: the revolution in lighting will not driven by the manufacturers of lightbulbs. To be sure, two of the Big Three—Osram Sylvania and Philips—have hedged their bets by investing in the LED chip-making subsidiaries, Osram Opto and LumiLEDs. The third lightbulb maker—GE—has dipped its toes in the water with a joint venture called GELcore. Its main claims to fame thus far are LED signs, traffic signals, and in-store refrigerator displays.* The Big Three are holding back because they know that if they target the market where they sell lightbulbs today, that would mean competing with their customers. No, the revolution will continue to be driven by zealous, impatient young men like Neal Hunter, Manuel Lynch,† George Mueller, Mark Schmidt, and Brent York. Will these revolutionaries have what it takes to build their businesses to global scale? Do not bet against them.

Other forces are working in favor of the rapid adoption of solid-state lighting. For example, the push for "green buildings," that is, combining good design and new technology to reduce the amount of energy that office blocks and other kinds of commercial edifices consume. Green buildings like New York's 7 World Trade Center use up to 50 percent less energy but command higher leasing fees, because companies want offices with a healthy work environment for their employees. New construction is not the only sector going green: for example, California boasts a thriving industry that specializes in refitting buildings with new technologies such as energy-efficient lighting systems. The green approach has already been formalized in various rating schemes such as the Leadership in Energy and Environmental Design standards developed by the US Green Building Council in 2000. Products such as Permlight's recessed can LEDs have been endorsed by the editors of *GreenSpec* and *Environmental Building News*. In 2005 the US tax law was amended to allow taxpayers to deduct the cost of energy-efficient appliances installed in commercial buildings.

The environmental impact of the solid-state revolution in lighting will be enormous. Look at a map of the world based on satellite photographs taken at night and you will see that light emissions from the United States, in particular the east coast of the country, are far higher

---

*On August 31, 2006, GE bought out its partner to take full ownership of GELcore, at the same time announcing a partnership with Nichia to help it make up lost ground on its rivals.

†But not at Permlight. In December 2006 Sea Gull Lighting Products, a leading supplier of lighting fixtures to major US home builders, announced that Manuel Lynch was joining the Riverside, New Jersey, based company as director of product technology. According to the press release, "Lynch will be responsible for greatly advancing the company's use of LED and emerging technologies."

than those from anywhere else. Most of this light is still produced by antiquated Edison-style incandescents. If every American household were to install energy-efficient lights in five of their most frequently used fixtures—the kitchen ceiling light, living room table and floor lamps, bathroom vanity, and outdoor porch lamp—the resultant drop in energy consumption would keep more than one *trillion* pounds of greenhouse gasses out of the atmosphere. That is equivalent to eliminating the pollution caused by more than eight million cars for an entire year, a $6-billion savings for householders equivalent to the annual output of more than twenty power stations.

How does Shuji Nakamura feel about what he has wrought? Out of all the consequences of his brilliant invention, what, I asked, gave him the most personal satisfaction? "Helping to prevent the effects of global warming," he replied, "and helping the people of third-world countries by giving them a safe lighting system."

POSTSCRIPT

On June 15, 2006, the very day I finished writing this book, it was announced that Shuji Nakamura had won the Millennium Technology Prize, the world's largest such award, for his invention of revolutionary new light sources. Presented biennially by a Finnish foundation, the prize is awarded "for a technological innovation that significantly improves the quality of human life." Nakamura is the second recipient of the prize, which was first awarded to Tim Berners-Lee, the developer of the World Wide Web. The award included a cash prize of 1 million euros (approximately $1.3 million). Shuji said that he hoped the award would raise awareness of the energy savings produced by using LEDs for illumination. He added that he planned to donate some of the prize money to groups that are helping to implement solid-state lighting in underdeveloped countries, like Dave Irvine-Halliday's Light Up the World Foundation. I cannot think of a worthier cause.

ACKNOWLEDGMENTS

There are two great pleasures in putting together a book such as this. One is the writing of the thing. The other is the opportunity to travel around the world and meet some truly exceptional individuals. Prime among them is, of course, Shuji Nakamura himself. I want to begin by thanking Shuji for agreeing to cooperate with this project, for his willingness to submit to my interviews and questions, both in person and via e-mail.

Shuji's colleague at UCSB Steve DenBaars was kind enough to write a letter in support of my proposal, as were Kevin Dowling, strategy and technology supremo at Color Kinetics, and Jo Ann McDonald, esteemed editor of *Compound Semiconductors Online*. Both Kevin and Jo Ann provided support in other ways, too. In addition to her lavish hospitality at the Legacy Ranch, Jo Ann introduced me to two key resources, Bob Karlicek and Warren Weeks, both of whom were kind enough to guide me through the finer points of the black art of crystal growth.

In my work on the history and technology of LEDs, gallium nitride devices in particular, I count myself extraordinarily fortunate in having been able to draw upon the wealth of experience that, after almost forty years in the field, Herb Maruska has accumulated. In addition to giving me personal tutorials on various aspects of light emitting devices, Herb also did me the honor of reading through first drafts of many of my chapters. Through the lonely months of writing, it was comforting to have such a good friend to rely on. Needless to say, all errors are attributable to my insufficient understanding of the subject.

If Herb knows pretty much everything there is to be known about the technology of LEDs, Bob Steele knows pretty much everything about the markets for solid-state lighting. Bob was generosity itself, and I am most grateful to him for his time.

Another of Jo Ann's friends, Robert Walker, kindly served as my guide to recent LED-related developments in China and Taiwan. Portions of chapter 15 first appeared as an article in *Forbes Asia*, and I am grateful for permission to reproduce them here.

Tim Whitaker, editor of *LEDs Magazine*, and his counterpart at *CompoundSemiconductor.net*, Michael Hatcher, are doing a magnificent job in chronicling the solid-state revolution in lighting. I have drawn much from their coverage for which I am beholden to them.

Many thanks also to my agent, Mike Hamilburg, for his conviction that my proposal was worthwhile, and for his perseverance in finding a home for it.

In addition, to Jesi Vasquez, administrative assistant extraordinaire in the Materials Department at UCSB, for being simultaneously so efficient and such a delight to deal with.

Thanks also to Joe Gramlich, for his thorough and meticulous copyediting.

Above all, I want to thank my mentor and friend Victor McElheny. Over many years Vic has provided unstinting advice, support, and encouragement, in this and my previous projects. I dedicate this book to him, with my deepest gratitude.

Melbourne, June 30, 2006

SOURCES

Page numbers for sources have been included when applicable.

INTRODUCTION

9. For information on reverse salients, see Thomas P. Hughes, *American Genesis: A Century of Invention and Technological Enthusiasm* (New York: Penguin Books, 1989), pp. 71–74.

10. Figures on overall benefits come from the US Department of Energy, *Illuminating the Challenges: Innovations in Solid State Lighting* (March 2004), a report based on a workshop held in November 2003. Available online at http://www.netl.doe.gov/ssl/PDFs/SSL_FinWorkSummary.pdf.

10. The US Department of Energy is currently tracking about 135 planned or proposed coal-fired power generation plants. See Senate Committee on Environment and Public Works, 109th Cong., February 9, 2006.

10. Herbert Kroemer's lemma on new technology, as quoted in an interview with John Vardelas of the IEEE History Center on February 12, 2003, is as follows: "The principal applications of any sufficiently new and innovative technology have always been and will continue to be applications created by that new technology."

11. Haitz's law was first made public in *The Case for a National Research Program on Semiconductor Lighting*, a paper delivered at the Optoelectronics Industry Development Association in Washington, DC, on October 6, 1999. In its original form, the law posited a 30 percent increase in light output per decade. See Roland Haitz, Fred Kish, Jeff Tsao, and Jeff Nelson, *The Case for a National Research Program on Semiconductor Lighting*, online at http://lighting.sandia.gov/lightingdocs/hpsnl_long.pdf.

13. Isaac Newton actually wrote "*ye* shoulders of giants" in a letter to Robert Hooke, February 15, 1676.

13. Mighty Atom, known as Tetsuwan Atom in Japan, aka Astro Boy.

14. Interview with Gerald Stringfellow, June 2, 1999.

14. Interview with Fernando Ponce, May 19, 1999.

14. Glenn Zorpette, "Blue Chip," *Scientific American*, August 2000.

15. Svante Linqvist, *Late Night Live*, ABC Radio National, December 5, 2005.

16. Report on gallium nitride, Hank Rodeen and Robert Steele, *GaN-Based Electronic Products, Applications, and Markets*, Strategies Unlimited, 2005.

16. Lester Eastman and Umesh Mishra, "Toughest Transistor Yet," *IEEE Spectrum* 39, no. 5, May 2002, pp. 28–33.

16. Interview with Umesh Mishra: October 20, 2005.

17. Nakamura quote from the foreword to Shuji Nakamura and Gerhard Fasol, *Blue Laser Diode: GaN Based Light Emitters and Lasers* (Berlin: Springer, 1997).

19. Shuji Nakamura, *Ikari no Bureikusuru* [Breakthrough with Anger] (Tokyo: Shueisha, 2001).

CHAPTER 1

This chapter, along with chapters 3 and 5, is largely based on Shuji Nakamura's autobiographical *Ikari no Bureikusuru* [Breakthrough with Anger] (Tokyo: Shueisha, 2001), referred to hereafter as *BWA*, and on my interviews with Shuji Nakamura: September 16, 1994; May 16, 1995; June 17, 1999; May 10, 2004; and October 20–21, 2005.

Interview with Nobuo Ogawa, May 16, 1995.

34. Description of Shikoku extracted from *Japan: An Illustrated Encyclopedia* (Tokyo: Kodansha, 1993).

CHAPTER 2

45. For Clarke's third law, see Arthur C. Clarke, *Profiles of the Future: An Inquiry into the Limits of the Possible* (New York: Harper & Row, 1961).

45. Interview with John Kaeding, October 20, 2005.

45. Interview with Warren Weeks, October 4, 2005.

46. Interview with Asif Khan, October 5, 2005.

45. Herb Maruska served as my guide for the technical explanations in this chapter.

For general background on LEDs and their history, see *inter alia*: Ken Werner, "Higher Visibility for LEDs," *IEEE Spectrum* 31, no. 7, July 1994, pp. 30–39; George Craford, Nick Holonyak, and Frederick A. Kish, "In Pursuit of the Ultimate Lamp," *Scientific American*, February 2001; Arpad Bergh, George Craford, Anil Duggal, and Roland Haitz,"Promise and Challenge of Solid-State Lighting," *Physics Today* (December 2001).

On Nick Holonyak, see the oral history interview conducted by Frederick Nebeker of the IEEE History Center, June 22, 1993; see also Tekla Perry, "Red Hot," *IEEE Spectrum* 40, no. 6, June 2003, pp. 26–29.

52. Interview with Nick Holonyak, April 5, 1994.

54. Interviews with George Craford, August 18, 1994, and May 14, 2004.

57. Oral history interview with Herbert Kroemer conducted by John Vardelas of the IEEE History Center on February 12, 2003.

CHAPTER 3

BWA and interviews with Shuji Nakamura, plus supplemental information on Nakamura from profiles such as Glenn Zorpette, "Blue Chip," *Scientific American*, August 2000.

71. Shuji Nakamura and Gerhard Fasol, *Blue Laser Diode: GaN Based Light Emitters and Lasers* (Berlin: Springer, 1997).

72. My attempt to describe how an MOCVD system works is based on a series of conversations, face-to-face, and via phone and e-mail with three experts: Herb Maruska, Bob Karlicek, and Warren Weeks.

76. Interview with Herb Maruska, November 9, 2005.

Interviews with Bob Karlicek, March 30, 2005, and May 10, 2005.

Interview with Warren Weeks, October 4, 2005.

78. Interview with Ramu Ramaswamy, November 30, 2005.

CHAPTER 4

The core reference for much of this chapter is Herbert Maruska, *A Brief History of GaN Blue Light-Emitting Diodes*, online at www.sslighting.net/lightimes/features/maruska_blue_led_history.pdf.

I was fortunate enough not only to interview Dr. Maruska, but also to follow up on the interview with questions over a period of months, all of which he was kind enough to respond to in great detail. Herb maintains a library containing copies of almost every paper published on the nitrides, items from which he very generously shared with me on many occasions.

A good introduction to the subject is the review article by F. A. Ponce and D. P. Bour, "Nitride-based Semiconductors for Blue and Green Light-Emitting Devices," *Nature* 386, no. 6623 (March 27, 1997): 351–59.

85. For a discussion on the blue light emitting materials see, *inter alia*, Robert Gunshor and Arto Nurmikko, "Blue-Laser CD Technology," *Scientific American*, July 1996.

86. Interview with Herbert Kroemer, October 21, 2005.

87. Interview with Jacques Pankove, October 29, 1994.

91. Edward Miller, "The Early History of Gallium Nitride Research" (undated memoir), online at www.compoundsemi.com/documents/articles/cldoc/5022.html.

Letter to Herb Marsuka, unpublished, November 19, 1971.

95. Interview with Isamu Akasaki, May 25, 1995.

Quotes from the transcripts of speeches, Isamu Akasaki, "Development of Gallium Nitride Blue Light Emitting Devices—Insight, Challenge, and Success," and Hiroshi Amano, "Passion and Struggle in the Development of Gallium Nitride Blue Light Emitting Devices," delivered at the Takeda Award Forum, Tokyo, November 2002, online at www.takeda-foundation.jp/en/award/takeda/2002.

Jeff Hecht, "Processes to Produce Blue LEDs Improve," source unknown but possibly *Lasers & Applications*, February 1992.

CHAPTER 5

BWA and interviews with Shuji Nakamura.

105. Francis Crick, *What Mad Pursuit: A Personal View of Scientific Discovery* (New York: Basic Books, 1988).

Shuji Nakamura, Yasuhiro Harada, and Masayuki Seno, "Novel Metalorganic Chemical Vapor Deposition System for GaN Growth," *Applied Physics Letters* 58, no. 18 (May 1991).

120. Glenn Zorpette, "True Blue," *Scientific American*, September 1997.

John Markoff, "A Race to Catch a Japanese Star on Blue Lasers," *New York Times*, October 27, 1997.

Bob Johnstone, "Out of the Blue," *Forbes Global*, September 6, 1999.

CHAPTER 6

Jo Ann McDonald, the esteemed editor of *LIGHTimes* and *Compound Semi News*, is sure that she wrote the original article in English on Nakamura's work in *Electronic Engineering Times*, but neither she nor I have been able track the details down.

132. For silicon carbide, see *inter alia*, "History and Current Status of Silicon Carbide Research" at the Purdue University Web site, www.ecn.purdue .edu/WBG/Introduction/exmatec.html.

132. Nakamura recounts his ignorance of stock options in *BWA*.

133. Background information about the Hunter brothers, Cree, the company's formation, and its early days comes largely from the following sources: Chris Chinnock, "Blue Laser, Bright Future," *Byte*, August 1995; "Engineering College Spawns Spin-off Successes," *Engineering News*, College of Engineering, North Carolina State University, October 9, 2000; Irwin Speizer, "In the Right Light," *Business North Carolina*, May 2003; Scott Warfield and Leo John, "Cree's $3 Billion Fight," *Triangle Business Journal*, June 20, 2003; Spencer Ante, "Blood Feud," *Business Week*, August 11, 2003.

Bob Johnstone, "Japan Enters US Triangle," *Far Eastern Economic Review*, May 31, 1990.

The figures on Cree's production and market share come from a letter Neal Hunter wrote to *Wired* magazine in June 1995 in response to an article I had written, "True Boo-roo," in the magazine's March edition on Shuji Nakamura's breakthrough and its implications. Hunter always felt that his company was not getting the credit it deserved in the press, and in not mentioning Cree in my article I had unintentionally stepped on his sore toe. "Bob Johnstone took a mental vacation when he wrote the somewhat fictional piece . . . in your March issue," Hunter fumed. However, he did end his rant by conceding that "the article has a few valid points, namely that Nichia made a significant development in blue LED brightness and that the future for the LED industry is tremendous."

"Cree Earmarks $300 Million for Expansion," *LEDs Magazine*, August 2004.

On LED Lighting Fixtures, see Leo John, "Cree Founder Lights Up New Firm," *Triangle Business Journal*, October 24, 2005; LED Lighting Fixtures press releases, online at www.ledlightingfixtures.com/press.htm.

CHAPTER 7

"Dialight Reports Zero Failures in High-power LED Traffic Lights," *LEDs Magazine*, October 11, 2005, online at http://ledsmagazine.com/articles/news/2/10/6/1.

149. Lynne Sagalyn, *Times Square Roulette: Remaking the City Icon* (Boston: MIT Press, 2002).

150. On autonomous vehicles, CMU, and the DARPA Grand Challenge, see, for example, John Markoff, "Robotic Vehicles Race, but Innovation Wins," *New York Times*, September 14, 2005.

156. Ken Shulman, "The Color of Money," *Metropolis*, May 2001.

Abigail Mieko Vargus, "Clayton Christensen and George Mueller Discuss the Innovator's Dilemma," *MIT Engineering Systems Division News Archives*, March 2003.

156. For an analysis of the implications of the Color Kinetics IP stakeout, see Tim Whitaker, "Patent Protagonists Head to Court," *LEDs Magazine*, April 2005.

Steven Cherry, "LEDs into Gold," *IEEE Spectrum* 42, no. 2, May 12, 2005.

161. Dietrich Neumann, *Architecture of the Night: The Illuminated Building* (Munich; New York: Prestel, 2002).

Andrew Blum, "A World of Light and Glass," *Business Week*, March 28, 2006.

"Roadside LED Matrix Responds to Traffic Flow," *LEDs Magazine*, May 2005.

165. For US artist James Clar, see Donnie Snow, "Technology Is Art," *Agence-France Presse*, February 27, 2006.

165. Michael Kimmelman, "To Be Enlightened, You Pull the Switch," *New York Times*, October 1, 2004.

On the impact of ambient light, see, for example, the report on the Public Interest Energy Research Project titled *Healthy Schools: Daylighting,*

Lighting, and Ventilation at the California Energy Commission Web site, www.energy.ca.gov; also, Luc van der Poel, "Ambient Experience: LEDs Soothe Hospital Patients," *LEDs Magazine*, July 2005.

CHAPTER 8

171. For background on Carmanah Technologies, see Ken MacQueen, "A Brighter Future," *Macleans*, December 1, 2003; Andrew Duffy, "The Serial Entrepreneur," *Victoria Times Colonist*, June 23, 2004.

See also Moneer Azzam, "The Case for Solar-Powered LED Lighting," *LEDs Magazine*, June 2005.

179. "Carmanah Awarded Contract for Solar-Powered LED Bus Stops," *LEDs Magazine*, September 26, 2005.

"Carmanah Ranked #37 in Canada's Fastest Growing Companies," *Profit*, June 5, 2006.

180. "Cyberlux Matches Attributes of LED Lighting to Market Needs," *LEDs Magazine*, October 2005.

CHAPTER 9

On the genesis of Light Up the World, see Bonny Munday, "Giving Light, Spreading Hope," *Reader's Digest Canada*, July 7, 2004; and "Light Up the World Taps LEDs for Villages Off-Grid," *Lighting.com*, January 19, 2004, online at http://www.lighting.com/content.cfm?id=112&page=/.

For background on energy usage in the third world, see Andrew Simms, "It's Time to Plug into Renewable Power," *New Scientist*, July 3, 2004.

CHAPTER 10

On the Khan group's breakthrough, see V. Adivarahan, A. Chitnis, J. P. Zhang, M. Shatalov, J. W. Wang, G. Simin, M. Asif Khan, R. Gaska, and M. S. Shur, "Ultraviolet Light Emitting Diodes at 340 nm Using Quaternary AlInGaN Multiple Quantum Wells, *Applied Physics Letters* 79, no. 25 (December 17, 2001): 4240–42.

202. On the DARPA SUVOS program, see Tim Whitaker, "SUVOS Pushes UV LEDs and Lasers to Shorter Wavelengths," *CompoundSemiconductor.net*, May 2004.

CHAPTER 11

214. The "Slave" anecdote and much of the other material in this chapter comes from *BWA*.

Brendan Buhler, "The Information Age's Father: UCSB's Herbert Kroemer," *Daily Nexus*, January 18, 2001.

Chiaki Kitada, "Blue about Japan," *J@pan Inc*, July 2001, online at http://www.japaninc.com/article.php?articleID=53.

Glenn Zorpette, "Blue Chip," *Scientific American*, August 2000.

CHAPTER 12

On the history of licensing technology in the semiconductor industry, see, for example, Committee on Japan, Office of Japan Affairs, Office of International Affairs, National Research Council, *US-Japan Strategic Alliances in the Semiconductor Industry* (Washington, DC: National Academy Press, 1992).

For a roundup of the initial patent battle, see "LED Patent Blues: The Dispute between Nichia and Toyoda Gosei," *CompoundSemiconductor.net*, May 4, 1999.

On Nakamura versus Nichia, see *inter alia* Waichi Tada, "Nichia Enrages Shuji Nakamura," *Nikkei Business*, May 28, 2001; Yoshiko Hara, "Researcher Sues Nichia over Blue LED Patent Rights," *Electronic Engineering Times*, August 24, 2001; Irene Kunii, "An Inventor Takes On Japan Inc.," *Business Week*, December 10, 2001; John Markoff, "A Rebel in Japan Is Hailed as an Innovator in US," *New York Times*, September 18, 2002; Ken Belson, "Japan Court Says Company, Not Inventor, Controls Patent," *New York Times*, September 20, 2002.

For the impact of the Nakamura case on other Japanese inventors, see Martyn Williams, "Toshiba Faces Claim from Flash Memory Inventor," *IDG News Service*, March 3, 2004; Norimitsu Onishi, "A Jolt to Team Japan: Bonus Demands," *New York Times*, May 1, 2005.

For Shuji on the Japanese education system, see, for example, Chiaki Kitada, "Blue about Japan," *J@pan Inc*, July 2001; Denis Normile, "Shuji Nakamura Speaks Out," *Science* 21, January 2005.

237. Tim Whitaker, "Nakamura Awarded $189 Million for LED Patents," *CompoundSemiconductor.net*, February 2, 2004.

238. Themis Editorial Department, *Blue Light Emitting Diode: Invented by Nichia Corporation and Its Young Engineers* (Tokyo: Themis, 2004).

Michael Hatcher, "Nichia Marches On, Nakamura Licks Wounds and Pays Lawyers," *CompoundSemiconductor.net*, March 2005.

CHAPTER 13

"Nakamura Hints at Improved GaN Structures," *CompoundSemiconductor.net*, September 22, 2005.

CHAPTER 14

262. *Republic of Cascadia* Web site is well worth a visit; online at http://www .zapatopi.net/cascadia/.

CHAPTER 15

271. For information on the Chinese Solid-State Program, see *LEDs Magazine*, February 2005. My information comes from a presentation made by the program's director, Wu Ling, at the Strategies in Light 2005 Conference presented by Strategies Unlimited, Burlingame, California, February 7–9, 2005.

"China's Burning Ambition," *Nature* 435 (June 30, 2005): 1152–54.

273. Keith Bradsher and David Barboza, "Clouds from Chinese Coal Cast a Long Shadow," *New York Times*, June 11, 2006.

Zhu Chengpei, "Dalien Develops Semiconductor Lighting Venture," *China Daily*, March 19, 2004.

David Barboza, "China Builds Its Dreams and Some Fear a Bubble," *New York Times*, October 18, 2005.

275. See, for example, "Environmental Protection among Top Six Tech Demand for Beijing Olympics," on the Beijing 2008 Web site. LEDs come under the heading of *Environmental Technologies*.

CHAPTER 16

279. Figures on energy usage from "Power Struggle," *Economist*, November 24, 2005.

For general information on fluorescents, see *inter alia* William Hamilton, "Incandescence, Yes. Fluorescence, We'll See," *New York Times*, January 7, 2007; "Lighting Efficiency Information," online at http://www.energy.ca .gov/efficiency/lighting.

CONCLUSION

297. Danit Lidor, "Times Square Ball to Get LED Makeover," *Forbes*, December 30, 2005.

297. "Lexus Debuts 2008 LS 600h L: World's First Full Hybrid V8 Luxury Sedan," *AutoMotoPortal*, April 13, 2006, online at http://www.automotoportal.com/ article/Lexus_Debuts_2008_LS_600h_L_Worlds_First_Full_Hybrid_V8_ Luxury_Sedan.

Information on Nichia target sales and employees from *LEDs Magazine* summary of article in *Nihon Keizai Shimbun*, April 5, 2005.

Yoshiko Hara, "Nichia Develops 100-lumens/W Efficiency White LED Chip," *Electronic Engineering Times*, March 13, 2006.

Reena Jana, "A Lamp to Light the Way," *Business Week*, May 19, 2006.

298. Bob Steele quote from Michael Hatcher, "Chip Makers and Lighting Specialists Get Switched On to New Possibilities," *CompoundSemiconductor.net*, April 2006.

299. "Philips Presents Aurelle LED Candles," *LEDs Magazine*, April 22, 2006.

300. "GE Takes Full Control of GELcore, Teams with Nichia," *LEDs Magazine*, online at http://ledsmagazine.com/articles/features/3/10/5/1.

"The Rise of the Green Building," *Economist*, December 4, 2004.

300. Robin Pogrebin, "7 World Trade Center and Hearst Building: New York's Test Cases for Environmentally Aware Office Towers," *New York Times*, April 16, 2006.

"BuildingGreen Announces 2005 Top-10 Products," *buildinggreen.com*, December 2005.

Richard Florida, "Map: The World Is Spiky," *Atlantic Monthly*, October 2005, pp. 48–49.

Matti Huuhtanen (AP), "Calif. Prof Awarded Millenium Tech Prize," *Mercury News*, June 15, 2006.

INDEX